Seismic Performance of Concre

Structures and Infrastructures Series

ISSN 1747-7735

Book Series Editor:

Dan M. Frangopol

Professor of Civil Engineering and
The Fazlur R. Khan Endowed Chair of Structural Engineering and Architecture
Department of Civil and Environmental Engineering
Center for Advanced Technology for Large Structural Systems (ATLSS Center)
Lehigh University
Bethlehem, PA, USA

Volume 9

Seismic Performance of Concrete Buildings

Liviu Crainic and Mihai Munteanu

CRC Press
Taylor & Francis Group
Boca Raton London New York Leiden

CRC Press is an imprint of the
Taylor & Francis Group, an **informa** business

A BALKEMA BOOK

Colophon

Book Series Editor:
Dan M. Frangopol

Volume Authors:
Liviu Crainic and Mihai Munteanu

Published by:
CRCPress/Balkema
P.O. Box 447, 2300 AK Leiden, The Netherlands
e-mail: Pub.NL@taylorandfrancis.com
www.crcpress.com – www.taylorandfrancis.com

ISBN 13: 978-0-367-44588-1 (pbk)
ISBN 13: 978-0-415-63186-0 (hbk)

Typeset by MPS Ltd, Chennai, India

Visit the Taylor & Francis Web site at
http://www.taylorandfrancis.com

and the CRC Press Web site at
http://www.crcpress.com

British Library Cataloguing in Publication Data
A catalogue record for this book is available from the British Library

Library of Congress Cataloging-in-Publication Data

Structures and Infrastructures Series: ISSN 1747-7735
Volume 9

Table of Contents

Editorial

Welcome to the Book Series *Structures and Infrastructures*.

Our knowledge to model, analyze, design, maintain, manage and predict the life-cycle performance of structures and infrastructures is continually growing. However, the complexity of these systems continues to increase and an integrated approach is necessary to understand the effect of technological, environmental, economical, social and political interactions on the life-cycle performance of engineering structures and infrastructures. In order to accomplish this, methods have to be developed to systematically analyze structure and infrastructure systems, and models have to be formulated for evaluating and comparing the risks and benefits associated with various alternatives. We must maximize the life-cycle benefits of these systems to serve the needs of our society by selecting the best balance of the safety, economy and sustainability requirements despite imperfect information and knowledge.

In recognition of the need for such methods and models, the aim of this Book Series is to present research, developments, and applications written by experts on the most advanced technologies for analyzing, predicting and optimizing the performance of structures and infrastructures such as buildings, bridges, dams, underground construction, offshore platforms, pipelines, naval vessels, ocean structures, nuclear power plants, and also airplanes, aerospace and automotive structures.

The scope of this Book Series covers the entire spectrum of structures and infrastructures. Thus it includes, but is not restricted to, mathematical modeling, computer and experimental methods, practical applications in the areas of assessment and evaluation, construction and design for durability, decision making, deterioration modeling and aging, failure analysis, field testing, structural health monitoring, financial planning, inspection and diagnostics, life-cycle analysis and prediction, loads, maintenance strategies, management systems, nondestructive testing, optimization of maintenance and management, specifications and codes, structural safety and reliability, system analysis, time-dependent performance, rehabilitation, repair, replacement, reliability and risk management, service life prediction, strengthening and whole life costing.

This Book Series is intended for an audience of researchers, practitioners, and students world-wide with a background in civil, aerospace, mechanical, marine and automotive engineering, as well as people working in infrastructure maintenance, monitoring, management and cost analysis of structures and infrastructures. Some volumes are monographs defining the current state of the art and/or practice in the field, and some are textbooks to be used in undergraduate (mostly seniors), graduate and postgraduate courses. This Book Series is affiliated to *Structure and Infrastructure Engineering* (http://www.informaworld.com/sie), an international peer-reviewed journal which is included in the Science Citation Index.

It is now up to you, authors, editors, and readers, to make *Structures and Infrastructures* a success.

Dan M. Frangopol
Book Series Editor

About the Book Series Editor

Dr. Dan M. Frangopol is the first holder of the Fazlur R. Khan Endowed Chair of Structural Engineering and Architecture at Lehigh University, Bethlehem, Pennsylvania, USA, and a Professor in the Department of Civil and Environmental Engineering at Lehigh University. He is also an Emeritus Professor of Civil Engineering at the University of Colorado at Boulder, USA, where he taught for more than two decades (1983–2006). Before joining the University of Colorado, he worked for four years (1979–1983) in structural design with A. Lipski Consulting Engineers in Brussels, Belgium. In 1976, he received his doctorate in Applied Sciences from the University of Liège, Belgium, and holds two honorary doctorates (Doctor Honoris Causa) from the Technical University of Civil Engineering in Bucharest, Romania, and the University of Liège, Belgium. He is an Honorary Professor of the Hong Kong Polytechnic University, of the Tongji, Southeast, Tianjin, and Chang'an Universities, and a Visiting Chair Professor at the National Taiwan University of Science and Technology.

Dan Frangopol is a Distinguished Member of the American Society of Civil Engineers (ASCE), Inaugural Fellow of the Structural Engineering Institute, Fellow of the American Concrete Institute (ACI), Fellow of the International Association for Bridge and Structural Engineering (IABSE), and Fellow of the International Society for Health Monitoring of Intelligent Infrastructures (ISHMII). He is also an Honorary Member of the Romanian Academy of Technical Sciences, Honorary Member of the Portuguese Association for Bridge Maintenance and Safety, and Honorary Member of the IABMAS-China Group. He is the initiator and organizer of the Fazlur R. Khan Distinguished Lecture Series (http://www.lehigh.edu/frkseries) at Lehigh University.

Dan Frangopol is an experienced researcher and consultant to industry and government agencies, both nationally and abroad. His main research interests are in the application of probabilistic concepts and methods to civil and marine engineering, including structural reliability, probability-based design and optimization of buildings, bridges and naval ships, structural health monitoring, life-cycle performance maintenance and management of structures and infrastructures under uncertainty, risk-based assessment and decision making, infrastructure resilience to disasters, and stochastic mechanics.

According to ASCE (2010) "Dan M. Frangopol is a preeminent authority in bridge safety and maintenance management, structural systems reliability, and life-cycle civil engineering. His contributions have defined much of the practice around design specifications, management methods, and optimization approaches. From the maintenance of deteriorated structures and the development of system redundancy factors to assessing the performance of long-span structures, Dr. Frangopol's research has not only saved time and money, but very likely also saved lives."

Dr. Frangopol's work has been funded by NSF, FHWA, NASA, ONR, WES, AFOSR and by numerous other agencies. He is the Founding President of the International Association for Bridge Maintenance and Safety (IABMAS, www.iabmas.org) and of the International Association for Life-Cycle Civil Engineering (IALCCE, www.ialcce.org), and Past Director of the Consortium on Advanced Life-Cycle Engineering for Sustainable Civil Environments (COALESCE). He is also the Chair of the Executive Board of the International Association for Structural Safety and Reliability (IASSAR, www.columbia.edu/cu/civileng/iassar), the former Vice-President of the International Society for Health Monitoring of Intelligent Infrastructures (ISHMII, www.ishmii.org), and the founder and current chair of the ASCE Technical Council on Life-Cycle Performance, Safety, Reliability and Risk of Structural Systems (http://content.seinstitute.org/committees/strucsafety.html).

Dan Frangopol is the recipient of several prestigious awards including the 2012 IALCCE Fazlur R. Khan Life-Cycle Civil Engineering Medal, the 2012 ASCE Arthur M. Wellington Prize, the 2012 IABMAS Senior Research Prize, the 2008 IALCCE Senior Award, the 2007 ASCE Ernest Howard Award, the 2006 IABSE OPAC Award, the 2006 Elsevier Munro Prize, the 2006 T. Y. Lin Medal, the 2005 ASCE Nathan M. Newmark Medal, the 2004 Kajima Research Award, the 2003 ASCE Moisseiff Award, the 2002 JSPS Fellowship Award for Research in Japan, the 2001 ASCE J. James R. Croes Medal, the 2001 IASSAR Research Prize, the 1998 and 2004 ASCE State-of-the-Art of Civil Engineering Award, and the 1996 Distinguished Probabilistic Methods Educator Award of the Society of Automotive Engineers (SAE). He has given plenary keynote lectures in numerous major conferences held in Asia, Australia, Europe, North America, South America, and Africa.

Dan Frangopol is the Founding Editor-in-Chief of *Structure and Infrastructure Engineering* (Taylor & Francis, www.informaworld.com/sie) an international peer-reviewed journal, which is included in the Science Citation Index. This journal is dedicated to recent advances in maintenance, management, and life-cycle performance of a wide range of structures and infrastructures. He is the author or co-author of more than 300 books, book chapters, and refereed journal articles, and over 500 papers in conference proceedings. He is also the editor or co-editor of more than 30 books published by ASCE, Balkema, CIMNE, CRC Press, Elsevier, McGraw-Hill, Taylor & Francis, and Thomas Telford and an editorial board member of several international journals. Additionally, he has chaired and organized several national and international structural engineering conferences and workshops.

Dan Frangopol has supervised 35 Ph.D. and 50 M.Sc. students. Many of his former students are professors at major universities in the United States, Asia, Europe, and South America, and several are prominent in professional practice and research laboratories.

For additional information on Dan M. Frangopol's activities, please visit http://www.lehigh.edu/~dmf206/.

Preface

This book arose from the need to pass on the life-long accumulation of knowledge in the field of reinforced concrete structures gained through the authors' teaching of under- and post-graduate students at the Technical University of Civil Engineering of Bucharest/Romania (TUCE) as well as through their experience as consulting engineers and structural designers.

The curricula of courses offered at the TUCE are substantially influenced by the fact that, being located at the intersection of three important tectonic plates, Romania is a country with high intensity seismic activity. The March 4th, 1977 destructive earthquake (Richter $M = 7.2$; 32 high-rise buildings collapsed in Bucharest, about 1600 people were killed and close to 12000 people were injured; two small towns – Zimnicea and Alexandria – were almost completely destroyed) was the biggest seismic event of the 20th century in Europe.

Taking into account this reality, most courses taught at TUCE place main focus on Earthquake Engineering topics. On the other hand, it is a tradition of TUCE to involve its faculty in structural design of earthquake-prone constructions and assessment and redesigning of earthquake-damaged ones. Those were important deciding factors in shaping the book's structure.

The book starts with a presentation of some fundamental aspects of reinforced concrete behavior quantified through constitutive laws for both monotonic and hysteretic loading. The basic concepts of post-elastic analysis such as plastic hinge, plastic length, fiber models, stable and unstable hysteretic behavior are defined and discussed accordingly. Based on the specific peculiarities of reinforced concrete behavior as quantified through its constitutive laws, two chapters (three and four) of the book present the analysis and design methods of concrete structures under static and seismic actions. The fundamental non-linear approach specific to these methods is highlighted and their specific use for concrete structures is exemplified through case studies.

Taking into account the seismic action peculiarities and specific features of seismic design philosophy as compared to the case of gravity-load-dominated structures, up-to-date seismic analysis methods as well as the modern seismic performance design approach are presented and discussed. In chapter five, the powerful concept of structural system is defined and systematically used to explain the seismic response as well as the procedures for analysis and detailing of common building structures. Three chapters (six, seven and eight) of the book have a pronounced applicative character. They treat the seismic behavior, analysis, design and detailing of frames, walls and dual systems. Case studies on the seismic design of each of the three types of concrete buildings are included. Chapter nine outlines some specific features of the concrete building behavior during major earthquakes. The book ends with the chapter "Concluding Remarks and Recommendations".

The book is not code-oriented. Rather than presenting code provisions, the volume tries to offer a coherent system of notions, concepts and methods which allow the understanding and implementation of any design code.

The content of this book is based mainly on the authors' personal experience resulting from their teaching and technical activities as professors and, at the same time, as consulting engineers and structural designers. Nevertheless, discussions and professional interaction with their colleagues from the Department of Reinforced Concrete at TUCE significantly contributed to the clarification of the topics and the final format of the book.

Many people were instrumental in helping the authors to realize this book project. Special thanks go to Dr. Mihai Pavel, Dr. Andrei Zybaczynski and Dr. Cristian Rusanu for drafting the case studies and providing several figures of the book. The photographs in chapter nine have been generously provided by Dr. Sever Georgescu from the Romanian National Research Institute for Constructions (INCERC). The authors are grateful to Mr. Vasile Hosu who performed the hard work of drawing the figures of the book and formatting the text.

The intellectually stimulating influence of Professor Dan Frangopol, editor of the book series "Structures and Infrastructures", is gratefully acknowledged especially for his valuable and constructive suggestions aimed at enhancing the book content and structure.

Last but not least, the authors want to thank their families for their love and unwavering support throughout this project.

This book aims to provide a powerful tool for both under- and post-graduate students as well as for structural designers, one that will enrich their knowledge and help them achieve a sound conception of and insight into seismic design of concrete buildings.

Liviu Crainic
Mihai Munteanu

Notations

Latin upper case letters

A	Area
A_c	Cross sectional area for concrete
A_s	Cross sectional area for reinforcement
A_{stirr}	Cross sectional area for stirrups
C_c	Compression in concrete for a cross section (internal force)
E_{el}	Elastic energy (recoverable)
E_c	Modulus of elasticity for concrete
E_{dp}	Dissipated energy
E_s	Modulus of elasticity for steel
EI	Stiffness of element
$E_c I_c$	Stiffness of a concrete element
F_b	Base shear force
F_{code}	Base shear force (code value)
F_{el}	Horizontal seismic force (elastic behavior)
G	Gravity type load
H	Horizontal load
H	Height
I	Second moment of area
L	Length
K	Stiffness
M	Bending moment
M_{cr}	Bending moment, cracking value
M_{el}	"Elastic" bending moment
M_p	Bending moment in the "plastic hinge"
M_u, M_{max}	Bending moment corresponding to failure
M_y	Bending moment at steel yielding initiation
M_E	Applied bending moment
M_{Ed}	Design value of the applied bending moment
M_G	Bending moment produced by gravity type loads
M_H	Bending moment produced by horizontal loads
M_R	Moment of resistance (capable bending moment)
ΣM_{Rb}	Sum of design values of moment of resistance for the beams framing in a joint

ΣM_{Rc}	Sum of design values of moment of resistance for the columns framing in a joint
M_s	Bending moment in seismic combinations of loads
N	Axial force
N_{Ed}	Design value of the applied axial force
P	Force
$S_1, S_2, S_3 \ldots$	Generalized forces
$S_e(T)$	Elastic response spectrum
$S_d(T)$	Design spectrum (elastic analysis)
T	Torsional moment
T	Natural vibration period
T_s	Tension in reinforcement for a cross section (internal force)
V	Shear force
V_{Ed}	Design value of the applied shear force
V_R	Shear force resistance value
V_{Rd}	Design value of the shear force resistance value
V_s	Shear force in seismic combinations of loads
W_t	Weight of masses

Latin lower case letters

a	Distance
a_g	Design ground acceleration
b	Width of a cross section
b_w	Width of the web on T, L, I cross sections (beams or walls)
d	Effective depth of a cross section
d	Diameter of a reinforcing bar
d_1	Concrete cover for the tensioned (less compressed) reinforcement layer
d_2	Concrete cover for the compressed (less tensioned) reinforcement layer
f_c	Compressive strength of the concrete
f_{cd}	Design value of compressive strength of the concrete
f_{ct}	Tensile strength of the concrete
f_{ctd}	Design value of tensile strength of the concrete
f_s	Tensile force in reinforcing steel
f_y	Yielding strength of the steel (tension or compression)
f_{yd}	Design value of yielding strength of the steel (tension or compression)
f_{ywd}	Design value of yielding strength of transverse reinforcement
g	Acceleration of gravity
h	Height or overall height of a cross section
h_b	Height of beam
h_c	Height of column
h_s	Distance between the reinforcing steel layers
k	Stiffness
k_{el}	Elastic stiffness
$[k], \{k\}$	Sectional stiffness matrix
k_M	Amplification factor (bending moment)
k_V	Amplification factor (shear force)
l	Length or span

l_0	Clear span or distance between two consecutive plastic hinges
m	Mass
m	Dimensionless value of the bending moment ($M/bd^2 f_c$)
n	Dimensionless value of the axial force
$p, p\%$	Reinforcing steel percentage (steel content $* 100$)
q	Behavior factor
r	Radius
t	Thickness
t	Time
$u, u(t)$	Ground displacement (incremental)
w	Width of the equivalent strut
$x, x(t)$	Relative displacement (incremental)
$y, y(t)$	Total displacement (incremental)
x	Neutral axis depth
z	Lever arm of internal forces

Greek upper case letters

Δ	Displacement
Δ_l	Elongation
$\Delta_1, \Delta_2, \ldots$	Generalized displacements
Φ_0	Plastic hinge overstrength factor
Ψ_0	Subsystem overstrength factor
Ω or ω	Overstrength factor (M_R/M_s)

Greek lower case letters

α	Angle or ratio
β	Angle or coefficient
γ	Angular deformation
γ	coefficient
γ_I	Importance factor
γ_{Rd}	Model uncertainty factor on design value of resistances
δ	Drift
δ_i	Interstory drift
ε	Equivalence factor
ε	Strain
ε_c	Strain in the concrete
$\varepsilon_c, \varepsilon_{c1}, \varepsilon_{c2}, \varepsilon_{c3}$	Compression strain in concrete at the peak stress f_c (model 1, 2 or 3)
$\varepsilon_{c,u}, \varepsilon_{c1,u}, \varepsilon_{c2,u}, \varepsilon_{c3,u}$	Ultimate compression strain in concrete (model 1, 2 or 3)
ε_t	Tensile strain
ε_s	Strain in the reinforcement
$\varepsilon_{s,u}$	Ultimate strain in the reinforcement
ε_y	Yielding strain in the reinforcement
θ	Rotation, bending angle
θ_p	Plastic rotation
λ	Proportionality coefficient (loading factor)

λ	Aspect ratio (l_0/h)
λ_0	Reinforcement overstrength factor
λ_p	Loading factor producing a plastic moment M_p
λ_u	Loading factor producing the failure of a structure (ultimate)
ϕ	Curvature $(1/r)$
ϕ	Virtual rotation
ϕ	Diameter of a reinforcing bar
ϕ_{av}	Average curvature
ϕ_p	Plastic curvature
ϕ_u	Ultimate curvature
η	Virtual displacement
μ	Steel content of a cross section (A_s/A_c)
μ_Φ	Sectional ductility factor (curvature)
μ_Δ	Sectional ductility factor (displacements)
μ_{nec}	Ductility demand (section or structure)
σ	Stress
σ_c	Stress in concrete
σ_s	Stress in the reinforcement steel
τ	Tangential stress
ω or Ω	Overstrength factor (M_R/M_s)

About the Authors

Liviu Crainic is Professor of Reinforced Concrete Structures at the Technical University of Civil Engineering Bucharest-Romania. He taught undergraduate and postgraduate courses for two years ('84/'85) at the National Polytechnic Institute of Alger (Algeria) as a visiting professor. Professor Crainic spent an academic year as research fellow at the Swiss Federal Institute of Technology (ETH) Zurich/Switzerland (1968–1969).

He earned his Ph.D. degree from the Technical University of Civil Engineering, Bucharest, Romania in 1974.

Professor Crainic has authored several books including *Energy Concepts and Methods in the Structural Dynamics* (co-author with S. Hangan, Romanian Academy Press, 1980), *Reinforced Concrete* (1993), *Reinforced Concrete Structures* (2003), and *Prestressed Concrete* (2007). He took part in drafting the *Romanian Code for Seismic Design of Buildings P100-92* and *Design of Reinforced Concrete Frame Structures* (NP 007/1997). He is author of several papers published in professional journals and of numerous reports presented to professional meetings.

In addition to the academic experience, his non-academic experience includes structural design work with direct contribution to the design of numerous reinforced concrete structures (multistory buildings, industrial buildings, water towers, tanks, silos) and to assessment and redesigning of earthquake-damaged reinforced concrete structures.

Professor Crainic is active in the Research Advisory Board of the Ministry of Public Works and Territory Planning. He is a member of the Task Group T3/3 of CEB "Assessment and Redesign of Existing Structures" and worked as a member of the Task Group 8 of the UNDRO Project RER/88/004 – "Earthquake Risk Reduction Network in the Balkan Region".

Professor Crainic is the Chairman of the Romanian Group of International Association of Bridges and Structural Engineering (IABSE), Member of the Romanian Committee of Concrete, Member of the Romanian Association of Civil Engineers, Member of the Romanian National Associations of Earthquake Engineering and Member of the Structural Design Engineers Association.

 **Mihai Munteanu** is Associate Professor of Reinforced Concrete Structures Department at the Technical University of Civil Engineering Bucharest-Romania.

He achieved his Degree in Civil Engineering, *Technical University of Civil Engineering of Bucharest* (B.S. and M.S. in 1984), followed in 1999 by a PhD degree for the thesis "Researches related to the behavior and computation of the structural-wall system structures in seismic areas".

Mihai Munteanu has a strong background in structural design, initiated between 1984 and 1990 when he was a Structural Designer with the *Research & Design Institute for Non-ferrous Metallurgy and Inorganic Chemistry Industry of Bucharest*, and continued, in the private sector, ever since.

Mihai Munteanu has more than 25 years of experience in the field of Structural Design Projects, Certified Checking and Technical Expert Review, in creating computer programs to structurally analyze and reinforce concrete sections and parts of structures.

Last but not least, he has been educating and counseling new generations of structural engineers for more than 25 years.

Mihai Munteanu is member of the Romanian Structural Design Engineers Association. He has co-authored several books and authored numerous articles that have been published both locally and internationally.

Chapter 1

Introduction

Abstract

This introductory chapter presents the basic notions and concepts of reinforced concrete structures for buildings. The elastic-plastic-viscous character of reinforced concrete behavior is highlighted. Types of schematization within structural modeling are discussed. Role and format of design codes are defined.

1.1 General

According to the general (classical) definition, *concrete* is an artificial stone obtained by hardening a mixture of *inert aggregates* (usually sand and gravel), *Portland cement* and *water*. According to the modern technologies, concrete could comprise also *additives* of different natures; the aggregates and the binding material can be different from these of the classical composition in order to ensure special performances: high mechanical resistance, high density or, by contrast, low density, etc. Nevertheless, the fundamental concrete properties are that of an *artificial stone*.

Like any other natural or artificial stone, concrete is characterized, from a mechanical point of view, by high compressive strength and much lower tensile strength.

Just like the stone-made works, *plain concrete* finds its usage in performing structural elements, or even structures subjected mainly to compression, like massive components of the foundations, bridge abutments, massive arches and domes, retaining walls, dams, etc.

The spectacular revolution in the use of concrete was realized, starting with the middle part of the nineteenth century, when the idea of combining concrete with *mild steel* (steel containing a low percentage of carbon) appeared. Basically, steel bars are included in fresh concrete expecting that, after hardening, the concrete resists the compressive stresses while the steel the tensile stresses. In this way a composite material is obtained – *reinforced concrete* – that has the advantages of both components i.e.: concrete and steel. Thus, reinforced concrete can be obtained in various shapes and sizes realizing a wide range of constructive forms; reinforced concrete is chemically more stable than other materials, requiring easy maintenance; from a mechanical point of view it is resistant so that it can be used in any conditions of loading, being able to resist tensile or compressive stresses behaving advantageously with bending, shear, torsion or any other combination of forces. Having such properties, reinforced concrete became, shortly after it was discovered, the most frequently used material for buildings' structures and for any other constructions.

The discovery, toward the end of the 1920's, of *prestressed concrete* was a remarkable application of the knowledge about concrete behavior achieved since then. It opened new perspectives for concrete use, especially in the domain of the elements with high structural performances: large spans, slenderness comparable to those of the steel constructions, etc. In prestressed concrete the low tensile resistance of the concrete is counteracted by introducing, before loading with the external forces, a system of compressive forces so that the tensile stresses developed by the simultaneous action of the compression and of the external forces, in each point of the element, has a limited magnitude. In fact, in prestressed concrete, the tensile stresses, generated by external actions, are carried through *decompression* of the elements' sections, subjected initially to compression by the prestressing force. If, under the simultaneous action of the pre-compressive force and any other combination of external forces, in all points of the element are developed only compression stresses it is said that the prestressing is *total*. If, in some loading combinations, tensile stresses appear of limited magnitude, resisted by ordinary reinforcement (un-prestressed), we say that the prestressing is *partial*.

Through a considerable effort of synthesis, developed in the last decades, a unique approach of plain concrete, reinforced concrete and pre-stressed concrete was realized, all these materials being treated under the name of *structural concrete*. In current language, the term "reinforced concrete" is actually used more frequently with the meaning of "structural concrete", because it represents the most important part of the constructions.

With every year, starting since it was discovered, the reinforced concrete has enlarged its use area. Today, a wide range of reinforced concrete *structural elements* are used like beams, columns, structural walls, foundations, plane or curved slabs of different forms, arches and vaults, trusses and many others. The structural concrete is used, on a large scale (sometimes exclusively) in bridges, in construction of silos and bunkers, in construction of tanks and water towers, for water-cooling towers and in the construction of many other structures. Practically, today, there are no constructions where *the reinforced concrete* is not present.

1.2 Behavior Peculiarities

Reinforced concrete is a composite material that results from the combination of two materials with different mechanical properties – concrete and steel.

Concrete behaves like an *elastic-viscous-plastic material*, with resistance and rigidity that differ in tension and compression. Compressed concrete behaves elastically within limited stresses domain (stresses less than 1/3–1/2 of the strength); for stresses that exceed this point, the concrete behaves like a pseudo-plastic material (when unloading, significant residual deformations occur due to the structural modification at microscopic scale). Subjected to long-term actions, concrete shows rheological (time-dependent or *viscous*) deformations, called *creep*.

The *mild steel* ("natural steel") is a typical elastic-plastic material having a linear-elastic domain, yielding point and a well defined yielding plateau.

In reinforced concrete, *at the interface concrete/steel*, at microscopic scale, complex phenomena are developed (elastic sliding, micro-fissures, local crushing) quantified, at macroscopic scale, by bond stresses variable with the loading step. All these phenomena strongly influence the cracking, the rigidity and, generally, the behavior of reinforced concrete elements loaded up to the failure.

Besides these aspects, resulting from the experimental and analytical study of reinforced concrete behavior, the engineer should take into account, even though not explicitly quantified, the existence of some *execution errors*, the *interaction* between structural and non-structural elements, the *in-steps execution history* of the reinforced concrete structures and other similar factors.

We may say that the reinforced concrete structures *have their own life*, which should be taken into account by the engineer when he selects the analytical models and when he decides on each detail of the structure.

Long term laboratory investigations, performed all over the world, sometimes through common resource programs of laboratories of different countries, as well as the lessons deduced from the *in-situ* performance of existing constructions, delivered an impressive quantity of information, (qualitative as well as quantitative), referring to the reinforced concrete structures behavior.

A big step forward, in this domain, was done with the development, starting with the 1960s, of the computer science (both hardware and software). It allowed to imagine and to realize complex computing models that simulate, with high degree of accuracy, the behavior of reinforced concrete under different types of loading.

Up-to-date design, research, as well as teaching, in the domain of reinforced concrete structures, are no more possible without taking into account these progresses.

1.3 Structural Modeling

The constructions involve *non-structural* elements (walls, partitions, finishing, casings), necessary to ensure its function and esthetics and *structural* components that ensure their *resistance* and *stability*.

We call *a structural system* the assembly of main structural components that resists and transfer actions (forces and imposed displacements/deformations) ensuring the resistance and the stability of the building. Generally speaking, the structural system is composed of three components: *(1) superstructure, (2) infrastructure and/or foundations* and *(3) the foundation soil*. Depending upon the type of construction these three components are specifically defined.

In order to simulate analytically the behavior of the structural systems under external actions, *the physical reality* (the elements as they are in reality, as well as the external actions) has to be replaced by *simplified schemes* or *models*, which take into account only certain characteristics of the real system, considered to be significant, and neglect others. We call this process *modeling for structural analysis*. For each structural system, the designer can choose simplified or more complex models, depending upon the nature of the investigated problem as well as the available resources (equipments, programs, trained people, etc.).

The structural modeling refers to the following aspects:

(a) *Geometrical schematization.* The structural elements are three-dimensional bodies. For calculation these bodies are generally replaced by geometrical abstractions like *linear members* (straight or curved) or *surface (2D) elements* (without thickness) plane or curved.

(b) *Physical schematization.* The structural elements, as well as the structure as a whole, are deformable bodies. The external loads deform the structure so that every point of the body receives *displacements* (linear displacements and rotations). Displacements represent *the kinematical response* of the structure to the external input. The magnitude of the response depends directly upon the deformability of the material of which the structure is made, upon the geometry of the structure and, obviously, upon the external load applied to the structure.

The relationship between *generalized forces* (forces and moments) applied at one *point* of the structure, at one *section* or on an *element* of the structure, and the corresponding *deformations*, represents an index of the deformability of the material, respectively of some segments of the structure. They are generically called *constitutive laws* (of material, of section, or of the element, as they refer to the material of which the structure is made, to a segment of unit length from a member or to an element of the structure). It is said also that the constitutive laws express the *physical aspect* of the structural analysis.

The basic constitutive law, from which all other constitutive laws are deduced, is that of the material, also named *characteristic curve*; it expresses the relation between the stresses and the corresponding strains, determined on a standard sample loaded in standard conditions until failure. The real stress-strain relationship (established by laboratory tests) is schematized for computation purposes by simplified curves. This operation is called *physical schematization.*

Following the physical schematization, the materials may be *elastic, elastic-plastic* or *rigid-plastic* (with negligible elastic deformations in respect to the plastic ones). If the characteristic curve depends significantly on the time that material has a *viscous* or *rheological* behavior.

(c) *Static schematization.* Being material bodies developed in real 3-D space, the structures have a distributed *mass* according to the dimensions and densities of the component elements. Acted by external loads (forces and/or imposed displacements), the points of the structure are moving with certain *velocities* and *accelerations* having magnitudes that depend on the way the loads are applied. If the accelerations are small so that multiplied with mass, in every point of the structure, gives inertial forces negligible in respect to the maximum external forces, we say that the structure is *statically* loaded and the corresponding structural analysis is *static.*

For performing static analysis, the real structure as well as the external loads is modeled within a *static scheme.*

The static scheme has to provide:

– Geometry and topology of the structure – the length of the elements (spans and height of the structure), their orientation (in plane and space) and the way in which these elements are interconnected;
– Structure supports – their nature (fixed, hinged, simple or elastic supports) and their position;
– Geometrical schematization for each element of the structure;
– Physical schematization for each element of the structure;
– Schematization of external loads: concentrated (pointed) or distributed forces, imposed displacements of the supports (caused, for example, by differential settlements) or of the elements (from temperature variations).

(d)   *Dynamic schematization.* When the external loads (forces and/or imposed displacements) induce within the structure displacements with accelerations which are no more negligible, the structural analysis has to take into consideration the *dynamic* character of the external action as well as of response (time-dependent inertial forces – and the time history of the response). This is the case of the structures subjected to seismic actions. In this case, the structural analysis has to be performed on the *dynamic model* of the structure.

The dynamic model provides:

– All specific data for static schematization;
– Dynamic degrees of freedom for the considered structure;
– Masses magnitude and distribution;
– Schematization of the external dynamic actions.

1.4 Design Codes

In order to ensure *an uniform, controlled level of safety* of reinforced concrete structures, the design process is performed according to specific legal documents, named *design codes.*

The codes specify the general design conditions ("requirements") and basic assumptions; they quantify the actions and the mechanical characteristics of the materials, state the computational methods and accepted simplifications, define the standard constructive details and, generally, offer all necessary data for designing according to a well defined, consistent, approach.

The design codes have a *technical*, as well as a *legal* meaning. The legal character of the design codes is well evidenced when there are to be established *responsibilities* in conflict situations. In such situations, the judgment criterion is based on the *strict compliance of the design with the principles and contents of the codes.*

Presently, there are *national* and *international codes.*

The national codes are specific to each country. They constitute a unique system involving components consistent to each other. It has to be known and correctly applied by designers.

The international codes have a *framework character*, having as main purpose to achieve a certain *compatibility* of the provisions of different national codes. Well known international codes for designing the reinforced concrete structures are the structural EUROCODES.

The design codes include usually *compulsory* requirements; however some provisions could have a *recommended* character explicitly specified within the code.

Through their specific format, the design codes use lapidary-type expressions, currently being left away explanations and technical or scientific justifications concerning the requirements.

It is the task of the user (designer or engineer) to apply the codes in full knowledge of their sense and of the theoretical bases of each requirement.

A major purpose of present work is to help students and structural designers to understand the fundamental aspects of behavior and analysis of reinforced concrete structures and, accordingly, to achieve knowledge for correct use of structural design codes and for sound conception and structural design.

1.5 Content of the Book

The present book comprises ten chapters.

First chapter "Introduction" briefly presents the general framework in which the book has been conceived. The basic concepts of *structural concrete, building structural system* and of concrete *structural models* are defined. They are the primary tools which allow explaining and understanding the structural response peculiarities of different types of concrete buildings under static and seismic actions. Meaning and purpose of design codes are highlighted.

Chapter two "Constitutive Laws" analyzes the up-to-date methods for modeling the response of reinforced concrete to external actions. The hierarchy of models quantifying the behavior peculiarities of reinforced concrete i.e. the *constitutive laws of different levels* is defined. Constitutive laws of reinforced concrete at *material, section* and *element* levels are thoroughly examined and their practical usage within analysis of concrete structures is suggested. Theoretical considerations are illustrated through some numerical examples.

The next two chapters of the book are devoted to the presentation of the reinforced concrete structures behavior peculiarities, as well as to their analysis and design methods under static and seismic actions. Analytical methods presented within the third chapter entitled "Behavior and Analysis of Reinforced Concrete Structures under Static Loads" are illustrated through numerical examples. Chapter four "Seismic Analysis and Design Methods for Reinforced Concrete Structures" highlights the two fundamental aspects of these methods as compared to the case of gravity-load-dominated ones: seismic action peculiarities and specific features of seismic design philosophy. Up-to-date seismic elastic and post-elastic analysis methods as well as the modern seismic performance design approach are presented and discussed. Within chapter five "Structural Systems for Multistory Buildings" the basic concept of *structural system* is defined and its importance in seismic design of different types of building structures is demonstrated. The next three chapters (chapters six, seven and eight) have a pronounced applicative character. They treat the seismic behavior, analysis, design and detailing of frames, walls and dual systems. Importance and rules for conceptual design of these types of structures are thoroughly discussed. Case studies for seismic design of each of three types of concrete buildings are included. The ninth chapter overviews some specific features of concrete building behavior during major earthquakes. Chapter ten presents "Concluding Remarks and Recommendations" synthetically drawn from the whole book contents.

Conclusions

Reinforced concrete is a material with complex structural behavior. Correct use of reinforced concrete structures for seismic prone buildings requires a deep understanding of its peculiarities as reflected in modern analysis and design methods as well as a specific technical sense in selecting advantageously structural solution. The major purpose of the present work is to help students and structural designers to deepen knowledge for sound conception and structural design of concrete buildings subjected to high intensity seismic actions.

Chapter 2

Constitutive Laws

Abstract

Structural behavior of reinforced concrete is quantified through relationships between generalized forces and corresponding displacements called constitutive laws. Within the present chapter three levels of constitutive laws are defined: material, section, and member. For each level monotonic as well as hysteretic loading are considered. Basic concepts which allow understanding and quantifying the reinforced concrete behavior peculiarities like: moment curvature relationship, plastic length, plastic hinge, spread plasticity, sectional and member ductility, energy dissipation, stable and unstable hysteretic behavior, are thoroughly examined. Up-to-date models describing the structural response of different types of R/C members, including short and long beams and columns, are presented. Numerical examples illustrating the theoretical concepts are included.

2.1 General Considerations

Structural elements, as well as the whole structures, are *deformable bodies*. This means that they respond to external loads by *deformations* (flexural, axial, shear and torsional) and *displacements* (linear displacements – like deflections – and rotational ones). Deformations and displacements generated by forces are called *kinematic response* of the structure (or element) to the external input. Different aspects of the structural response are lumped together under the notion of *structural behavior*.

Quantitatively, the structural behavior is expressed through relationships between generalized *forces* and corresponding *displacements*.

Relationships between the generalized forces and the corresponding deformations are called *constitutive laws*.

The background for determining constitutive laws for structural members is the study of mechanical properties of the material (or *materials*) of which the structure is made.

The mechanical properties (or mechanical *behavior*) of a material is expressed through relationships between stresses and strains developed in a standard specimen loaded according to standard procedures. In the US the detailed description of testing *procedures* and *specimens*, for different materials (concrete, steel, aluminum, wood, etc.), are given in ASTM (American Society for Testing and Materials) specifications. Stress-strain relationship for a given material is called the *material constitutive law*.

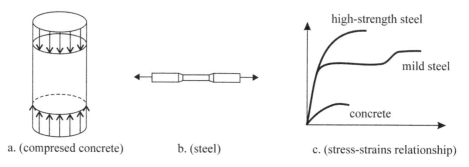

a. (compresed concrete)   b. (steel)   c. (stress-strains relationship)

Figure 2.1 Constitutive laws for different materials

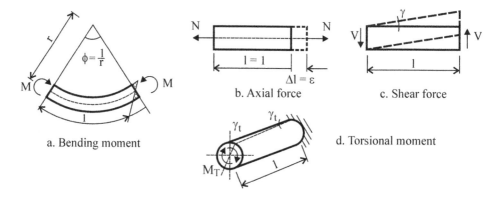

a. Bending moment   b. Axial force   c. Shear force   d. Torsional moment

Figure 2.2 Definition of sectional constitutive laws

Accordingly, at *material level*, the generalized forces are *stresses* and the generalized displacements are *strains*.

Many structures are made of structural elements that can be *geometrically* modeled as *one-dimensional members*: beams, columns, and arches. The mechanical equivalent of total stresses acting on each section of the member is the *internal force*: bending moment, axial force, shear force and torsional moment (or torque).

Each internal force generates a specific deformation:

Bending moment → curvature (angular deformation)
Axial force → linear deformation (elongation/shortening)
Shear force → sliding
Torsional moment → twist (angle).

Relationships between internal forces and corresponding deformations are called *sectional constitutive laws*. Sectional constitutive laws quantify the structural behavior of element segments having a length equal to one.

Deformations of each elementary segment, integrated along the element length, generate *relative displacements* of the member extremities – linear and rotational.

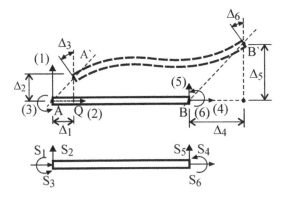

Figure 2.3 Member deformations

Table 2.1 Levels of constitutive laws

Level of constitutive law	Generalized forces	Generalized displacements
1. Material (concrete, steel)	Stresses	Strains
2. Section	Internal forces (M, N, V, Mt)	Deformations (curvature, axial, shearing and twisting deformations)
3. Member	End joints forces (vector)	End joints displacements
4. Structure	External forces (vector)	Structure displacements

Relationships between internal forces acting at the ends of a member and the corresponding displacements are the *member constitutive laws*. They express (quantify) the structural behavior of the whole member, detached from a structure (Crainic, L. 2003).

The same approach could be extended for the whole structure. The relationships between the external forces acting on a structure and the corresponding displacements could be considered as "constitutive laws" of the structure. Actually, determination of the response of the whole structure to external forces is a more complicated task than that of sections and elements. It is performed through structural analysis. So, we speak about constitutive laws only at material, section and element level.

Summarizing, we say that the constitutive laws at different levels are expressed through relationships between the quantities defined in the Table 2.1. The constitutive law at each level is based on constitutive laws at a previous one.

The above definitions and considerations are particularly useful for the study and understanding of the structural behavior of reinforced concrete.

Reinforced concrete is a composite material, resulting from the association of two different materials – steel and concrete – each of them having a specific behavior under loads.

Laboratory investigations evidenced that the concrete strength depends upon the nature and sign of stresses and its deformability is nonlinear and time-dependent. We call the concrete a *viscous-elastic-plastic* material.

Reinforcing steel behaves as an *elastic-plastic* material having a ductility that varies according to steel grade and fabrication procedure.

Bond between concrete and steel is variable too according to the duration and nature of loading (static or dynamic), stress state of bars' surrounding concrete etc.

Reinforced concrete structural response is a result of these factors. In order to predict the structural behavior of reinforced concrete structures, the constitutive laws at section and member level have to be investigated.

Because of the non-linear character of reinforced concrete behavior, constitutive laws should be examined for all loads' combinations. For evident practical reasons, at each level, only some loads' combinations will be considered, namely those that are most frequent and significant for R/C structures' behavior.

2.2 Constitutive Laws for Reinforced Concrete Components

2.2.1 Concrete

Theoretically, the constitutive laws of concrete have to be expressed as three-dimensional functions: stress-strain-time. Commonly, a simplified approach considers separately (a) time-dependent effects and (b) quasi-instantaneous loading accepting, globally, that concrete constitutive law is expressed by stress-strain relationship under short-term loading.

Time-dependent phenomena namely deformations under long-term constant stresses (creep), behavior under reversal loads, concrete fatigue and hysteretic behavior ("low-cycle fatigue") are separately treated.

Current structural analysis (static) is based upon stress-strain relationships determined for short-term loads.

Dynamic analysis, especially for seismic actions, takes into account the *structural hysteretic behavior*.

These aspects of concrete behavior will be briefly examined below.

Stress-Strain Relationships. Plain concrete behaves like an elastic-plastic material with limited ductility (Fig. 2.4). In compression, after peak stress value reached for a strain magnitude of about 0.002, independent of concrete strength, a certain *residual strength capacity* was evidenced for concrete under *constant strain rate loading*.

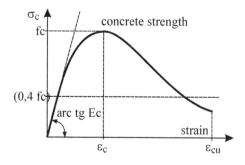

Figure 2.4 Stress-strain relationship for compressed concrete

Ultimate strain in compression for plain concrete is about 0.003–0.004.

For practical purposes, codes accept simplified relationships for stress-strain curve.

The most used simplified curve in compression is a parabola for concrete strain up to 0.002 (2‰), continued with constant stresses for strains between 0.002 and 0.0035 (3.5‰), which is considered to be the ultimate compressive strain of unconfined (plain) concrete (Fig. 2.5).

Concrete ductility in compression can be improved through confining by spiral or transverse rectangular reinforcement (hoops) (Fig. 2.6, 2.7 and 2.8).

Tensioned concrete has a brittle failure (Fig. 2.9).

Hysteretic behavior. Loading and unloading of a concrete specimen up to failure shows, for strains exceeding 0.002, significant resistance and stiffness degradation (Fig. 2.10).

Peak stresses envelope is the stress-strain curve for monotonic loading ("skeleton curve").

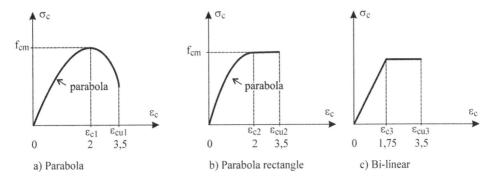

a) Parabola   b) Parabola rectangle   c) Bi-linear

Figure 2.5 Simplified relationship for stress-strain curve

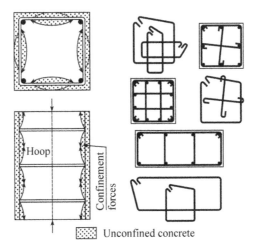

Unconfined concrete

Figure 2.6 Concrete confined with rectangular hoops

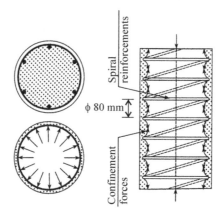

Figure 2.7 Concrete confined with spiral reinforcement

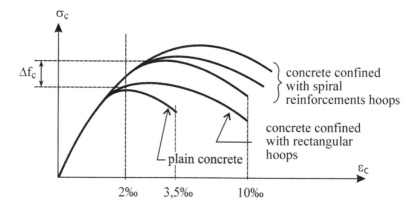

Figure 2.8 Concrete confined with rectangular hoops and spiral reinforcement

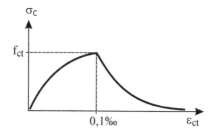

Figure 2.9 Stress-strain relationship for tensioned concrete

For analytical purposes one can find in the literature several stress-strain relationships, both for monotonic or hysteretic loadings, for plain or confined concrete. In Figure 2.11 is shown, for example a type of constitutive law for concrete.

For analytical purposes different simplified curves were accepted as, for example, that shown in Figure 2.11.

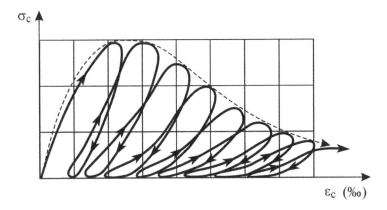

Figure 2.10 Hysteretic behavior of concrete (Park, R. & Paulay, T. 1975)

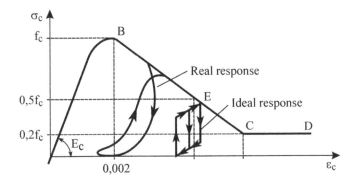

Figure 2.11 Simplified model for hysteretic behavior of compressed concrete

2.2.2 Reinforcing Steel

Stress-Strain Relationships for Monotonic Loading. For reinforced concrete structures two steel types are used: low-carbon, hot-rolled smooth or deformed bars and strain-hardened steel, especially smooth cold-drawn wires (usually available as fabric made welded wire meshes).

Typical stress-strain curves for two steel types are shown in Figure 2.12.

Analytical models accept simplified stress-strain curves for steel as shown in Figure 2.13 (Menegotto, M. & Pinto, P.E. 1973).

Natural steel has considerable ductility; however, for modeling reinforced concrete member behavior, ultimate steel strain is limited to magnitudes between 1% and 5% (0.01 to 0.05), depending upon the analysis approach.

Hysteretic Behavior. Loading/unloading of steel specimens for strains within post-elastic range lead to stress-strain curves as shown in Figure 2.14. These

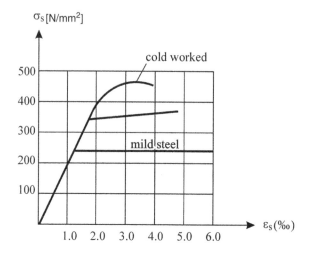

Figure 2.12 Simplified stress-strain curves for different steel grades

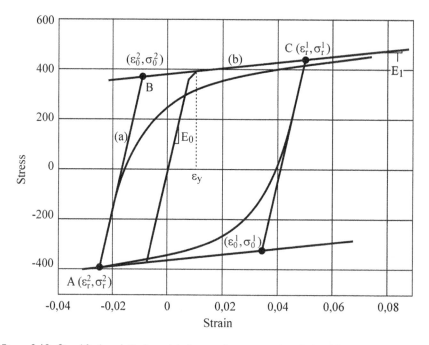

Figure 2.13 Simplified analytical models for steel stress-strain relationships

curves evidence the *Bauschinger-effect*, i.e. modification of stress-strain curve shape with loading reversal (especially decreasing of apparent yield limit has to be noticed).

Simplified stress-strain curves for reversal loading are shown in Figure 2.14.

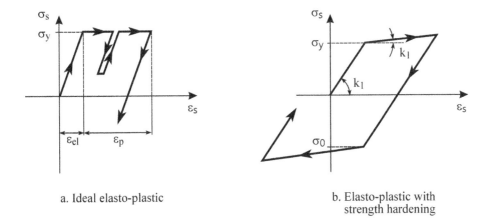

a. Ideal elasto-plastic   b. Elasto-plastic with
strength hardening

Figure 2.14 Stress-strain curve for reversal loading

2.3 Constitutive Laws for Reinforced Concrete Sections

2.3.1 *General*

Sectional constitutive law refers to behavior of a segment of a structural element, having the length equal to one, and internal forces constant along the element.

At both segment ends act internal forces with same magnitude generating corresponding deformations.

Sectional constitutive law is the relationship between these forces and corresponding deformations for *monotonically increasing loading* up to failure, or for different kinds of *repeated loads* (for instance: alternative loading in elastic range, or alternative loading within post-elastic range, or repeated loading between certain limits, etc.).

The most general sectional constitutive law could be written in matrix format as:

$$\{S\} = [k]\{\Delta\} \tag{2.1}$$

where $\{S\}$ is the internal forces (generalized forces) vector,
$\quad\{\Delta\}$ the deformations vector and
$\quad[k]$ the sectional stiffness matrix.

Because of non-linear behavior of reinforced concrete, sectional stiffness matrix terms are not constant but *functions of loading level*. For the same reason, the superposition principle is no more valid, so that [k] should be determined for each forces combination and for each loading history.

For practical purposes, a simplified approach is adopted. Accordingly, a principal (predominant) internal force is chosen as variable parameter expressing sectional behavior, while other forces are considered *secondary effects*. Sectional constitutive law becomes, thus, a relationship between two quantities: the principal internal force and the corresponding deformation.

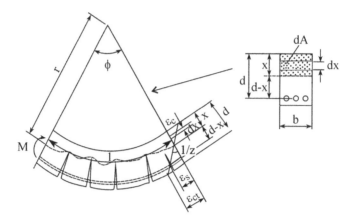

Figure 2.15 Curvature of a cracked reinforced concrete element

Usually, the most significant internal force for describing sectional behavior is the *bending moment*. Consequently, the sectional constitutive law is expressed as *moment-curvature relationship*. The effect of *axial forces* is afterward examined, taking into account constant magnitudes of these forces in connection with variable bending moments.

2.3.2 Moment-Curvature Relationships for Sections with Pure Bending

We consider a segment of a reinforced concrete straight member having a length of one unit subjected to *planar pure bending*. The bending moment increases gradually from zero ($M = 0$) up to its ultimate magnitude ($M = M_u$), corresponding to the section failure.

According to the moment magnitude, the element is deformed and passes through six *behavior stages*: (1) elastic non-cracked (2) crack initiation (3) elastic cracked (4) initiation of steel yielding (5) post-elastic behavior (6) ultimate stage through concrete failure (crushing) or rebars failure.

The rotation angle of two end sections of the segment is the specific deformation generated by bending moment. Because the bending moment is constant along the member, deformed shape is circular and the rotation angle between end sections is the *curvature* of deformed member.

Quantitatively, the sectional behavior is described by a relationship between moment and the corresponding curvature. The slope of moment/curvature curve is the *sectional stiffness* of the member.

A more detailed analysis reveals that, after cracking, neutral axis depth is variable (Fig. 2.15).

Current models accept a constant *mean depth of the neutral axis* along the member and, accordingly, mean magnitudes of concrete and steel stresses and strains.

Because of symmetry, the plane section assumption (Bernoulli's assumption) is valid, so that the following geometrical equations can be written:

$$\phi = \frac{\varepsilon_c}{d} = \frac{\varepsilon_s}{d-x} = \frac{\varepsilon_c + \varepsilon_s}{d} \tag{2.2}$$

The material constitutive laws (stress-strain curves) for both concrete and steel and above geometrical equations allow determining stress distributions throughout the cross section and, then, internal forces (bending moment) for each behavior stage. For a given maximal compressive concrete strain (which defines the loading level) the only unknown quantity is the neutral axis depth x which can be determined by expressing the equilibrium of internal forces and stresses:

$$\iint \sigma \, dA = 0 \tag{2.3}$$

Thereafter, the bending moment corresponding to selected loading level is determined by writing moment equations of stresses about an axis perpendicular to bending plane:

$$M = \iint \sigma \cdot n \, dA \tag{2.4}$$

Practical approach is a step-by-step trial and error procedure implemented within computer programs.

The main steps are the following:

(i) Input data:
   – Section geometrical characteristics b, h, d, As,
   – Stress-strain curves for concrete (compression, tension) and for steel
   – Number of desired points to determine moment-curvature relationship i

(ii) Step number j – initially j = 1 then
      j = j + 1

(iii) Choose maximum compressive concrete strain as a fraction of ultimate one:

(iv) Choose (arbitrary) neutral axis depth x

(v) Determine reinforcement strain and maximum concrete tensile strain.
   Remark: Maximum concrete tensile strain shall be under its ultimate magnitude. Cross section portion with elongation over ultimate concrete tensile strain is cracked. Thus, check-up of maximum elongation shows if member is cracked or not.

(vi) Determine concrete and steel stress at each level of the cross section, taking into account the linear strains distribution (according to Bernoulli's assumption) and stress-strain curves of both materials.

(vii) Calculate total compressive and tensile stresses and check if eq. (2.3) is fulfilled. If the answer is:
   – NO – go to step (iv);
   – YES – continue.

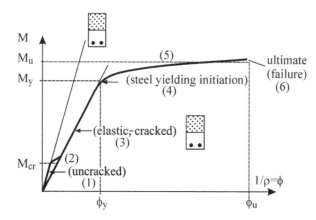

Figure 2.16 Typical moment-curvature relationship for reinforced concrete sections with pure bending

(viii)   Determine the curvature with eq. (2.2) and corresponding moment with eq. (2.4). Save the results.

(ix)   Go to step (ii)

Numerous computer programs exist for calculation of moment-curvature relationship according to above described procedure or to equivalent ones.

Comments on the results

Computer programs for determining moment-curvature relationship allowed performing intensive investigations about the influence of various parameters on reinforced concrete sections behavior. Typical moment-curvature curve is shown in Figure 2.16.

Influence of steel and concrete stress-strain relationships, of steel amount, of section shape, etc. on this curve was examined. Several results of such investigations are shown in Figure 2.17 (Hangan, S. & Crainic, L. 1980). It can be noticed that the sections subjected to pure bending, reinforced with mild (ductile) steel in moderate amounts, show a ductile behavior.

Difference between moment magnitude corresponding to stage (4) (steel yielding initiation) M_y and that corresponding to failure (ultimate moment) M_u is as much as 10 to 15% of M_y. Curve portion corresponding to un-cracked behavior (stage 1) can be neglected, taking into account also the possibility of pre-cracking.

Consequently, moment-curvature relationship can be modeled, for practical purposes, as bilinear curve with limited plateau length (Fig. 2.18). Area below the curve is the energy absorbed by the member segment loaded up to failure. It has two components:

– a recoverable part, corresponding to elastic deformations, and
– a "lost" part (transformed into heat) called *dissipated energy*.

The capacity of a section (or, better said, of a segment having a length of one) to develop post-elastic deformations and, thus, to dissipate energy can be quantified

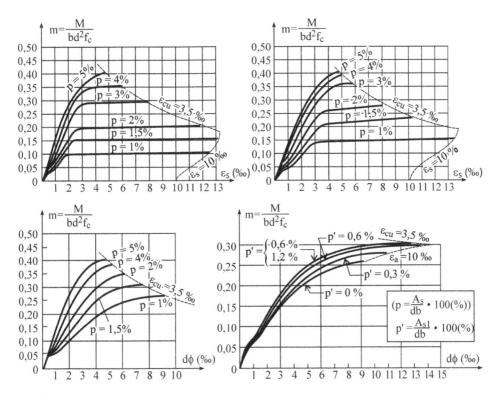

Figure 2.17 Moment-curvature relationship for different steel amount

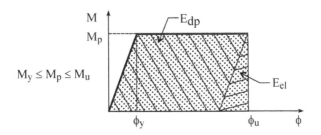

Figure 2.18 Simplified moment-curvature curve

through sectional ductility factor, defined as the ratio between ultimate curvature and that corresponding to yielding:

$$\mu_\phi = \frac{\phi_u}{\phi_y} \tag{2.5}$$

For bilinear moment-curvature relationship, as shown in Figure 2.18, energy dissipating capacity of a section is proportional to ductility factor.

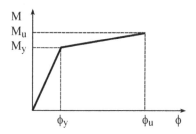

Figure 2.19 Simplified shape for section with significant difference between M_y and M_u

When difference between yielding moment M_y and the ultimate one M_u is significant for structural analysis simplified shape of moment-curvature relationship as shown in Figure 2.19 is adopted.

2.3.3 Moment-Curvature Relationships for Eccentrically Compressed Members

For members with bending and axial force, two internal forces (axial force and bending moment) act, in each section, generating two kinds of deformations (axial – elongation or shortening – and curvature). Consequently, constitutive law should be described through a 4D surface (M, N, curvature, and axial deformation).

Simplified approach accepts that flexural effect is predominant (e.g. bending moment and curvature) and axial force remains constant during loading until the member failure. It is also accepted that no significant mutual influence exists between curvature and deformation generated by axial force.

Accordingly, moment-curvature relationship for eccentrically compressed members can be determined through the same procedure as for those with pure bending. The only difference lies in the equation for determining neutral axis depth (equilibrium condition), which becomes:

$$\iint \sigma \, dA = N \tag{2.6}$$

Generally, dimensionless quantities are implemented rather than absolute magnitudes of forces and geometrical parameters, namely:

Dimensionless magnitude of bending moment

$$m = \frac{M}{bd^2 f_c} \tag{2.7}$$

Axial force intensity

$$n = \frac{N}{bd f_c} \tag{2.8}$$

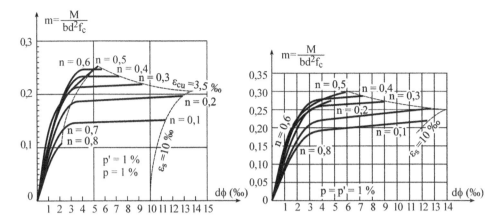

Figure 2.20 Constitutive laws for different axial force intensities

Dimensionless curvature

$$d\Phi = \varepsilon_s + \varepsilon_c \qquad (2.9)$$

Comments on the Results

Typical moment-curvature relationships, for different axial force intensities, are shown in Figure 2.20.

These curves evidence that axial force magnitude has a strong influence on eccentrically compressed sections behavior. Globally, following specific behavior paths it can be stated that:

–   For small axial force intensities ($n = 0.0$ up to 0.2) the behavior is close to that of sections with pure bending, the failure being *ductile*;
–   For moderate axial forces (n between 0.3 and 0.5) the failure is governed by compressed concrete being of *semi-ductile* type (limited ultimate curvature). Sectional ductility can be improved, in this case, through confining concrete that allows it to develop substantial ultimate strains;
–   Sections with high magnitude of axial force (n exceeding 0.5) have a *brittle* failure due to generalized concrete crushing. The only possibility to improve ductility, in such cases, is to increase cross section size and/or concrete strength.

2.3.4 Moment-Curvature Relationships for Sections Subjected to Reversal Loading

Structural static or dynamic analysis to seismic actions requires information about behavior of sections subjected to repeated loads within post-elastic range (hysteretic behavior). Both analytic and experimental investigations were performed in this field.

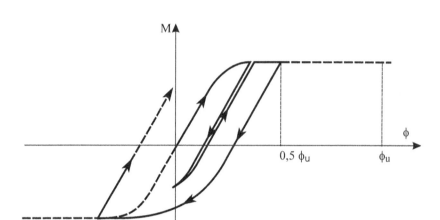

Figure 2.21   Stable behavior

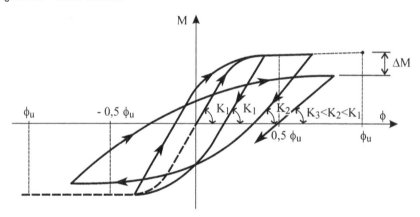

Figure 2.22   Unstable behavior

Most significant results can be synthesized as follows:

For limited deformations magnitudes hysteretic moment-curvature relationship is *stable*, e.g. the shape of the curve remains unchanged for more cycles of reversal loads (Fig. 2.21);

For maximal deformations close to ultimate ones *unstable hysteretic behavior* was evidenced, e.g. at each loading cycle (loading-unloading – reloading in opposite sense-unloading) decrease of maximal moment ("resistance degradation") and/or stiffness ("stiffness degradation") were found (Fig. 2.22).

Resistance and stiffness degradation is generated by progressive section damage due to concrete crushing, steel post-elastic buckling, etc.

Globally, it can be stated that stable hysteretic behavior occurs when maximum curvature doesn't exceed the half of ultimate one:

$$\phi_{max} \leq 0.5\phi_u \tag{2.10}$$

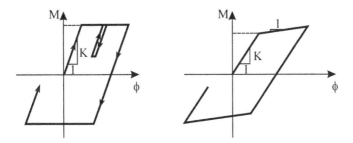

Figure 2.23 Simplified hysteretic moment-curvature relationships – stable behavior

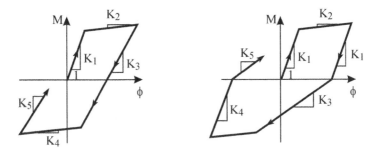

Figure 2.24 Simplified hysteretic moment-curvature relationships – unstable behavior

In the literature, one can find a large number of analytical constitutive laws for R/C sections subjected to reversal loading. These are to be used in various computer programs dealing with concentrated inelasticity models (see paragraph 2.4.7).

Hysteretic constitutive laws are shown in Figure 2.23 – stable behavior and Figure 2.24 – unstable or degrading behavior (Hangan, S. & Crainic, L. 1980).

2.4 Constitutive Laws for Reinforced Concrete Members

2.4.1 *General Considerations*

The portion of a structure between two consecutive nodes is called *element*. Very simple, statically determinate structures can be also considered as elements.

Generally, an element is loaded with *internal forces* acting on its extremities, which are reactions of the rest of the structure, and *external forces*. Mechanical behavior of the elements is expressed through relationships between *forces* and corresponding *deformations*. These relationships are called *element constitutive laws*. The element constitutive laws are generally written as set of algebraic equations, since an element is acted by more than one force (external and/or internal).

The displacement in a point of the element is the sum of displacements of elementary segments situated on both sides of the considered point. The sectional constitutive law quantifies the behavior of each elementary segment, having the length equal to one. Therefore, the element constitutive laws are deduced from the constitutive laws of the element sections.

The constitutive laws of reinforced concrete sections are non-linear relationships (see chapter 2.3). So, the constitutive laws for reinforced concrete elements will be expressed through algebraic set of equations with coefficients that depends on the loading level and the "loading history" (the way in which loading was applied).

The most general constitutive laws for reinforced concrete elements can be condensed in a matrix form. Stiffness matrix for reinforced concrete elements depends on sectional stress-strain relationship, on loading level and on loading history.

The structural analysis through stiffness matrix with variable components depending on loading level is the most exact method but it requires important resources (memory space and time for processing) and the physical phenomena developed within element are hard to follow.

2.4.2 Types of R/C Members According to Their Behavior up to the Failure

We refer to linear reinforced concrete members – beams or columns – subjected to loading (forces or node displacements) similar to that developed under seismic actions.

In a first step, global behavior under monotonic in-plane loading of a linear element up to the failure is considered. Their behavior is significantly influenced by the magnitude of shear force developed within the element as compared with that of bending (with or without axial force).

Referring to the ultimate state, the type of failure depends upon the ratio between maximum normal stress generated by bending moment M within the element σ_x, and the maximum shear stress τ_{xy} due to the shear force V. Within the general assumptions of elasticity, σ_x/τ_{xy} is proportional with the length of *shear arm* (or *shear span*) (Fig. 2.25):

$$M/Vh = H/h \qquad\qquad (2.11)$$

where: h is the cross section height and H – distance between the inflexion point of deformed element (i.e. section with zero moment) and the end cross section.

For a fixed ends element like in (Fig. 2.26) instead of shear arm can be considered its *aspect ratio* $L/h = 2H/h$ designated sometimes also as (geometrical) *slenderness*.

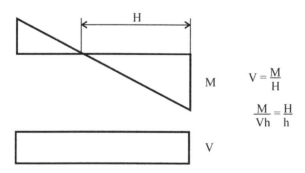

Figure 2.25 Definition of the shear arm

Experimentally it has been found that three situations can be encountered:

(a) Failure governed by *bending*
(b) Elements with *bending/shear failure*
(c) Elements showing a *shear failure*

Bending failure occurs for slender elements having an aspect ratio $L/h > \sim 4$–5. Normally, it has a ductile character (starts with steel yielding and finishes by compressed concrete crushing).

Bending/shear failure starts (for elements correctly designed and detailed so that shear capacity is greater than the flexural one) by flexural cracks developed within the zones of maximum moment. From flexural cracks (normal to the element axis) inclined shear cracks are initiated. Along a diagonal critical crack the final failure occurs, splitting the element. This type of failure is semi-ductile.

Shear failure is typical for short elements having an aspect ratio $L/h < \sim 2$–3. This is a violent, explosive failure. Compressive axial force accelerates the shear failure of the short columns through a "sledge effect" i.e. tendency of sliding of the upper part of the element on the inclined crack under the effect of vertical (axial) force (Fig. 2.27).

These statements can be summarized as shown in Figure 2.28.

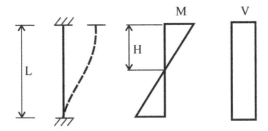

Figure 2.26 Shear arm for a fixed ends element

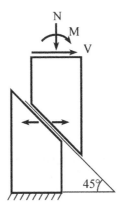

Figure 2.27 Failure of a short column

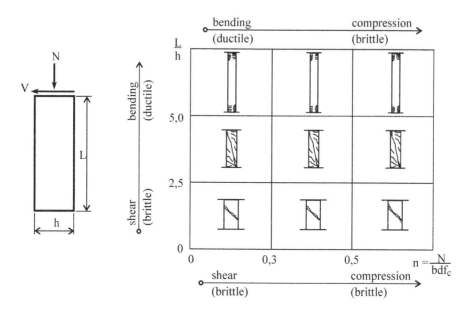

Figure 2.28 Types of failure for elements with different aspect ratios and axial force

2.4.3 Force-Deflection Relationship for a Reinforced Concrete Cantilever. Notion of "Plastic Hinge" (elements with "concentrated" inelasticity)

A simplified approach, useful for a better understanding of the post-elastic behavior of reinforced concrete members, considers simple elements, such as the cantilever.

Let us consider a cantilevered element loaded with a transverse force P on the top extremity, having a magnitude that gradually increases up to the element failure.

Constitutive law of this element is expressed through the relationship between force P and deflection Δ.

The cantilever is a statically determined system, so both internal forces and deflections can be easily determined. One has:

$$M_x = P \cdot (l - x) \tag{2.12}$$

$$M_{max} = P \cdot l \tag{2.13}$$

The cantilever behavior results from the behavior of all its cross sections (actually of each elementary part with length equal to one). Section behavior is described by moment-curvature relationships.

We assume that the moment-curvature relationship is known for each section of the element. Since the moment diagram is known for each loading step (see the previous equations) it is easy to determine the curvature distribution along the element.

For a given magnitude of the external force, the cantilever sections can be in elastic domain if $M_x < M_y$, or in post-elastic domain if $M_y < M_x < M_u$, where M_x is the

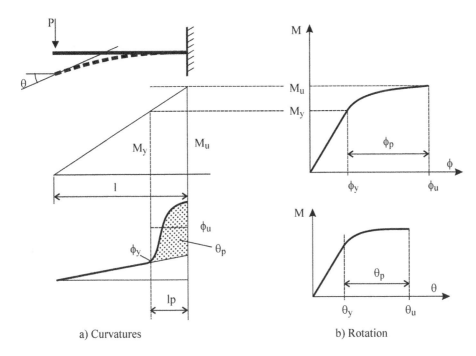

a) Curvatures   b) Rotation

Figure 2.29 Definition of plastic rotation

bending moment in point x along the element, M_y is the yielding of the flexural reinforcement and M_u is the ultimate (failure) bending moment value (Fig. 2.29).

As long as all sections remain in elastic domain relationship between force P and deflection Δ is linear:

$$P = k_{el}\Delta \left(\text{if } P < \frac{M_y}{l} \right) \tag{2.14}$$

where k_{el} represents the member *elastic stiffness*:

$$k_{el} = \frac{3 \cdot (EI)}{l^3} \tag{2.15}$$

and (EI) is the *sectional elastic stiffness* (the slope of moment-curvature curve).

When the external force P increases over the magnitude corresponding to the yielding of the base section

$$P \geq \frac{M_y}{l} \tag{2.16}$$

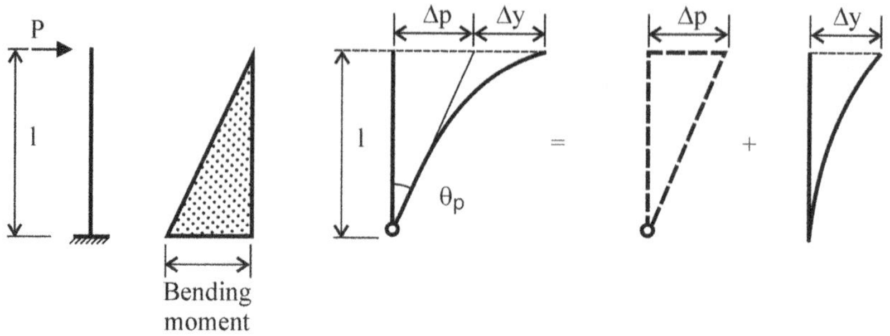

Figure 2.30 Displacements of a cantilever accepting the plastic hinge assumption

over a certain length l_p ("plastic length") plastic deformations will be developed. Theoretically, the plastic length is spread over the portion between base section and the section with $M_x = M_y$:

$$l_p = \frac{M_{\max} - M_y}{M_{\max}} \cdot l \tag{2.17}$$

When the external force magnitude further increases, at limit, in the base section the ultimate moment M_u is reached.

Notion of "Plastic Hinge"

For practical purposes, it is more convenient to consider that all plastic deformations are concentrated in a section situated at the centroid of the plastic deformation diagram. Due to plastic deformations, in this section, the element is suddenly bent similar to the case of a hinge. On the adjacent sections of this "hinge" a pair of moments M_p act, having an average magnitude between M_y and M_u:

$$M_y \leq M_p \leq M_u \tag{2.18}$$

We called this section "*plastic hinge*". This notion is not a physical reality but a simplified model aimed to simplify the structural analysis.

The section in which a potential plastic hinge appear, behaves elastically as far as $M_x < M_y$. When $M_x = M_y$, this section can no more transfer moments with magnitudes that overpass M_p and the element axis is breaking.

For reinforced concrete elements, the presence of the normal and inclined cracks, the local slip of reinforcing bars as well as many other factors, lead to a curvature distribution which differs considerably from that resulted theoretically.

Normally, the plastic length is longer than $l_P = ((M_{\max} - M_y)/M_{\max}) \cdot l$ and, accordingly, the plastic rotations have greater magnitude.

The elastic-plastic deformation can be considered to consist of two components: an elastic deformation and a rigid body rotation about the plastic hinge with an angle called *plastic rotation* θ_p.

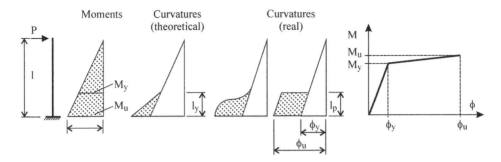

Figure 2.31 Rotation of plastic hinge

It is usually accepted that the rotation in plastic hinge can be determined considering a constant distribution of curvatures along the plastic zone:

$$\theta_p = [\phi_{av} - \phi_y] \cdot l_p \tag{2.19}$$

Generally, we assume that:

$$\phi_{av} \approx (\phi)_u \tag{2.20}$$

Consequently

$$\theta_p = [\phi_u - \phi_y] \cdot l_p \tag{2.21}$$

The approximations involved by this equation can be observed in figure 2.31.

A plastic hinge is defined, from the point of view of structural analysis, by three parameters:

Plastic length l_p

This parameter has to cover several phenomena that are not taken into account in above considerations, especially the effect of inclined cracks due to shear forces. Generally, it is accepted $l_p = (0.8 - 1.2)d$, where d is the effective depth of the cross section.

Plastic rotational capacity θ_p

This parameter is the maximum magnitude of the rotation angle, which can be developed by a plastic hinge. The following values give global information about the rotational capacity of some reinforced concrete elements:

7/1000–10/1000 (radian) for current columns and beams
4/1000–5/1000 (radian) for structural walls

More exact and differentiate values have to be taken into account for different analysis purposes. For example, FEMA 356 (November 2000) recommends specific rotational capacities to be used in analysis associated to different seismic performances for building rehabilitation.

Plastic moment M_p

This parameter represents the nominal magnitude of the constant bending moment supposed to be developed within the plastic hinge. The plastic moment ranges between M_y and M_u.

2.4.4   Elements with Distributed Inelasticity

Instead of accepting the simplified approach of plastic deformations concentrated in plastic hinge, advanced modeling of member behavior which takes into account the spread of inelastic deformations have been developed. By this way, many specific features of the seismic post-elastic response of reinforced concrete members could be understood and quantified like: difference between force-controlled and displacement-controlled behavior, 2D and 3D flexural behavior, hysteretic behavior, etc.

For instance, a cantilevered column loaded with a lateral monotonic force at its top has a force-displacement relationship up to the failure like in Figure 2.32a. If the same element is loaded with incremental monotonic imposed displacement at its top, up to the failure, the force-displacement diagram is shown in Figure 2.32b. In case a) the "failure" is defined by the load capacity of the member (when diagram starts to descend). In case b) a "descending branch" is evidenced since the failure is defined through the maximum element displacement. This means that, after the force capacity is reached (through concrete crushing), a residual force capacity still exists due to presence of flexural reinforcements and of (damaged) concrete residual capacity (strain softening in Fig. 2.11).

More information about spread plasticity and fiber models will be given in paragraph 2.4.7.

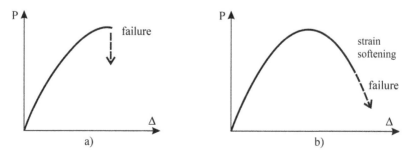

Figure 2.32 P-Δ relationship for a cantilever loaded a) by monotonic forces, b) by monotonic displacements

2.4.5 *Behavior of Elements with High Shear*

Let us consider a reinforced concrete member working in linear elastic range.

As shown in Figure 2.25 and 2.26, M/Vh is the shear span ratio. The lower the shear span ratio is, the greater the influence of the shear stress will be (Marti, P. 1991).

On the other hand, for the *short-type* members i.e. members with an aspect-ratio L/h lower than 2–3 (see Fig. 2.28) each end of the element is a *discontinuity* region (de Saint-Venant principle). Practically the entire element is a discontinuity region. Therefore the elastic field is two-dimensional and has to be analyzed only in two-dimensions. For the discontinuity regions the "beam theory" (Bernoulli's assumption of linear distribution of strains) did not apply (Schleich J. 1991).

The analysis of such a region will be made by using the strut & tie method, an application of the truss analogy (Fig. 2.33) (Anderheggen, E. & Schleich, J. 1990).

The resisting internal mechanism is formed by a compressed diagonal strut and by two tensioned ties (the flexural reinforcement).

The failure occurs in one of these two manners: a) flexural steel yields in tension (the ties) or, b) the diagonal strut fails (brittle) in compression.

Note that in the shown simplified model the transverse reinforcement is not modeled. Its role is "only" to confine the diagonal concrete strut.

For *short-type* and medium sized elements, i.e. members with an aspect-ratio L/h in a range below 5, more elaborated truss-resisting internal mechanisms can be developed, taking into consideration the apport of the transverse reinforcement too (see Fig. 2.34).

For the member model, a part of the shear is transferred from one end to the other through two compressed struts and a tensioned tie (stirrup) as shown in Figure 2.34a) (the path A-B-C-D). For the whole element and the whole shear intensity this mechanism is leading to a hyperstatic truss, having two families of compressed fan-type

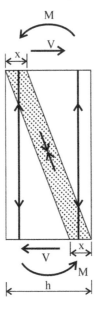

Figure 2.33 Short-type member. Truss analogy

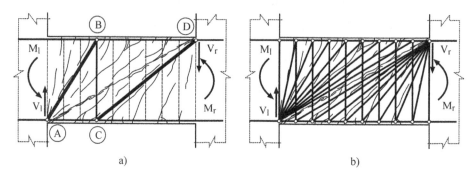

Figure 2.34 Truss analogy considering the transverse reinforcement. a) transfer of a part of the shear force from a margin to other margin. b) the whole truss mechanism

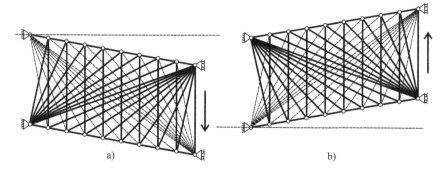

Figure 2.35 Truss analogy for alternate cycles. a) loading, b) unloading

struts radiating from the opposite corners, a number of tensioned ties modeling the stirrups and two tensioned upper and bottom chords for the flexural reinforcement layers (Fig. 2.34b).

The failure is reached by developing of the following mechanisms: a) crushing of the concrete in the compressed struts, b) yielding in tension of the flexural steel reinforcement, c) yielding of the stirrups, or by a combination of these three effects. The hyperstatic truss is transformed into a kinematic mechanism (Crainic, L. & Munteanu, M. 2003).

For cyclic actions, two families of fan-type struts will be developed alternatively in opposite corners of the element (Fig. 2.35). The families once compressed will become tensioned for alternate cycles (and will crack in tension) and vice versa. This phenomenon reduces dramatically the compressive strength of the struts.

Moreover, the cracks will intersect stirrups which locally will yield. Because of the remnant deformations in steel, these cracks will not close totally for small reversal loading. All these phenomena lead to a drop of member stiffness during unloading and small reversal loadings quantified through pinching effects and very narrow loops (reduced energy dissipation) (Fig. 2.36).

The overall aspect of the loops is a "double S" or "inverted S" shape.

More detailed aspects of this truss models will be discussed in paragraph 7.3.3. Coupling Beams' Behavior.

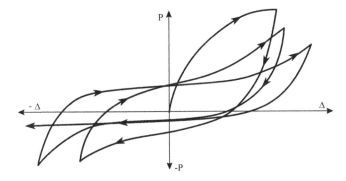

Figure 2.36 Hysteretic relationships for member with high shear

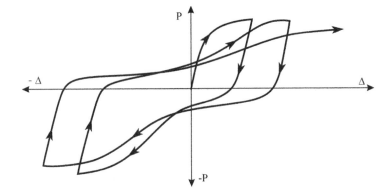

Figure 2.37 Pinching effect for members with large incursions within post-elastic range

Note that for this type of element the *"member's constitutive laws"* refers to the entire element which can no longer be studied section by section.

2.4.6 *Elements Subjected to Reversal Loading*

We consider a cantilever loaded at its extremity. This is considered to be the typical element for studying the behavior under repeated loads.

Its behavior depends upon the shear/flexure ratio as defined for members monotonically loaded (see paragraph 2.4.2).

For elements predominantly subjected to bending (elements with an aspect ratio $l/h > 5$) the constitutive law, expressed through relationship force/displacement at the end section, is similar to the hysteretic moment/curvature relationship (paragraph 2.3.4).

However, members with large incursions in the post-elastic range, close to the failure, show *pinching* of the hysteretic response (Fig. 2.37). Accordingly, energy dissipation capacity of these elements is diminished. The explanation is the following: Within first loading, flexural cracks (perpendicular to element axis) occur. When the external force changes the sign, the previously compressed concrete becomes (gradually) tensioned and, at a certain loading level, will crack. At the same time the cracks in previously compressed zone are still open due to residual (plastic) deformation of the reinforcement. So, the cracks on one side merge with cracks on the other. Consequently, the shear force is carried, in certain sections, only by the dowel effect, and the element stiffness suddenly decreases. For higher magnitude of external force, the previously opened cracks (now in the compressed zone) will close, and the element stiffness increases. In this process secondary effects occur too, which precipitates the element damage: progressive deterioration of the compressed concrete, gradual pull out of tensioned reinforcing bars and buckling of compressed longitudinal steel.

For elements predominantly subjected to high shear (elements with an aspect ratio $L/h < 2$–3) the pinching effect of the hysteretic response is still more visible (Fig. 2.36). For these members, the inelasticity phenomena and the failure mechanism are spread along the whole length of the element. Cyclic loading develops cracks spread along and across the two diagonals diminishing the concrete capacity to resist in compression, as shown in paragraph 2.4.4. Moreover, yielding of the stirrups in zones intersected by diagonal cracks leads to larger remnant deformations and greater diminishing of the element stiffness.

This is why, for these elements, the force/displacement constitutive law in the end sections is no longer likely the moment/curvature relationship in paragraph 2.3.4 but becomes a "double S" type diagram, showing very large sliding effects if the load path direction changes.

2.4.7 Computer Models for R/C Members

We call *element* any portion of a structure between two consecutive nodes. The element could be loaded or not with external forces.

For elastic elements the presence of external forces along the element is not relevant, since their effect could be treated separately according to the superposition principle.

For elastic-plastic domain, the overall behavior of the element depends upon all forces acting on the element and, also, upon the loading history. Accordingly, all effects have to be taken into account for each loading step.

Elastic elements

We consider an element with length l loaded with internal forces M, N, V at its ends. The directions on which the forces act and on which displacements are developed are called cinematic degrees of freedom. For planar loaded elements there are 6 degrees of freedom (D1, D2,..., D6), three on each node.

Relationships between node forces and corresponding displacements can be written as follows:

$$S_1 = k_{11} \cdot \Delta_1 + k_{12} \cdot \Delta_2 + \cdots + k_{16} \cdot \Delta_6$$
$$S_2 = k_{21} \cdot \Delta_1 + k_{22} \cdot \Delta_2 + \cdots + k_{26} \cdot \Delta_6$$
$$\vdots \qquad\qquad\qquad\qquad\qquad\qquad (2.23)$$
$$S_6 = k_{61} \cdot \Delta_1 + k_{62} \cdot \Delta_2 + \cdots + k_{66} \cdot \Delta_6$$

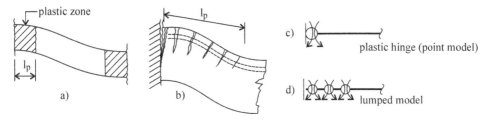

Figure 2.38 Beam model with end-concentrated inelasticity; a) Overall behavior; b) Detail; c) Point-hinge model; d) Lumped plasticity model

and in matrix form:

$$\{S\} = [k][\Delta] \tag{2.24}$$

where:

$\{S\}$ – the force (acting on the element ends) vector
$\{\Delta\}$ – the displacements vector
$[k]$ – the stiffness matrix.

For the elastic domain, the stiffness matrix components are constants, easy to be determined, as follows.

A displacement of magnitude equal to one is successively imposed to each node (on the direction of each degree of freedom) while all others are blocked.

From the above equations one has:

$$S_j = k_{ij} \quad (i, j = 1, \dots, 6) \tag{2.25}$$

So, k_{ij} represent the reaction at the end i when a displacement equal to one is applied at j end and all other displacements are blocked.

Reinforced concrete elements under service stage could be considered to behave elastically.

Elements with post-elastic behavior

Concentrated inelasticity models ("point-hinge" and "lumped" plasticity models).
We accept that the plastic deformations are concentrated either in plastic hinges or along a "plastic zone" situated at the member ends (Fig 2.38).

The easiest way to model the member post-elastic behavior is by using plastic hinge assumption (see 2.3.3). According to this assumption, when (in a critical section) plastic moment M_p is reached, a plastic hinge occurs i.e. a hinge on which, at its two faces, the constant moment M_p acts. The rest of the element remains elastic.

The structures subjected to seismic actions have critical sections usually situated on the member ends. Post-elastic stiffness matrix of an element is, thus, determined through progressive loading, step by step, with imposed proportional displacements at its ends. As long as the moments in critical sections don't reach the plastic moment M_p,

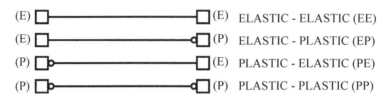

Figure 2.39 Element stiffness matrix corresponding to different locations of plastic hinges in
a beam

the element behaves elastically and its stiffness matrix is that of an elastic element $[k_{EE}]$.
When at one element end (i.e. critical section) $M = M_p$ a plastic hinge is considered to
occur at that section. For the next loading stages, the moment in plastic hinge remains
constant and the element keeps behaving elastically. It is advantageous to consider,
for next loading steps, only the loading increment. For that, plastic section behaves
like a true hinge (with no moment). The stiffness matrix corresponding to incremental
loading is called tangential. Tangential stiffness matrix is computed as for an elastic
element having a normal hinge at the end in which plastic moment was previously
reached. At each incremental loading step the tangential stiffness is determined through
similar procedure. Consequently, at different loading levels, the member ij could be in
one of the states of Figure 2.39 and its elastic-plastic tangential matrix is:

$$[k] = [k_{EE}] \quad \text{if } Mi < Mp,i \text{ and } Mj < Mp,j$$
$$= [k_{EP}] \quad \text{if } Mi < Mp,i \text{ and } Mj = Mp,j$$
$$= [k_{PE}] \quad \text{if } Mi = Mp,i \text{ and } Mj < Mp,j$$
$$= [k_{PP}] \quad \text{if } Mi = Mp,i \text{ and } Mj = Mp,j$$

So the post-elastic stiffness matrix depends on loading level and loading history.

For the lumped plasticity models the plastic component of the end node displacement
is the sum of the plastic deformations along the plastic zones. In such cases a simplified method of summation of the effects in the three zones to determine the member
assembly behavior can be used.

In most frequent case we consider that the forces increase proportionally with a
single parameter. Typically the axial-flexural coupling is neglected. We accept also that
the displacements at each end are independent to the displacements of the other end.
This means that the hysteretic behavior of the plastic zones will be computed apriori
(usually by a different computer program) before beginning the structural analysis of
the building.

Elements with distributed inelasticity ("slice" and "fiber" models)

The members are modeled as a whole inelastic element. The inelastic behavior is defined
at the section level. We define along the member a number of control-sections, and
the behavior of the assembly is obtained by integrating onto the length of the element
(Fig. 2.40). Usually the Bernoulli's assumption is assumed to be correct in these points.

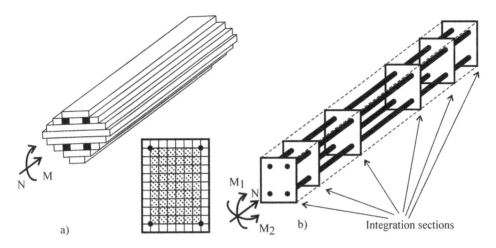

Figure 2.40 Beam model with distributed inelasticity; a) Slice-model; b) Fiber model

Practically, for uniaxial bending and axial force the member is partitioned in a number of "slices" or "layers" and for biaxial bending and axial force the partitioning will be made in "fibers" having the length of the element. For each of these layers or fibers is assumed a type of uni-dimensional hysteretic constitutive law (i.e. plain concrete, confined concrete, steel reinforcement – see Fig. 2.11 and 2.14). By including the inelastic hysteretic model into the material-type constitutive law, the post-elastic behavior of the member assembly is assured (Spacone, E. & Filippou, F.C. & Taucer, F.F. 1996).

In comparison to the concentrated inelasticity models, theoretically, the nonlinearity can take place in any section of the element and can "spread" to a neighborhood section during the loading-unloading cycles. In practice, because we integrate the effects only in the control-sections only these points are monitorised.

Clearly, the greater the number of layers or fibers and the greater the number of control-sections, the more accurate the final results will be.

Of course, both models shown above have advantages and disadvantages as follows.

The use of concentrated inelasticity models is leading to faster and less storage consumption of the intermediate data computation. On the other hand, assuming an a priori length for the "plastic zone" and a priori behavior relationships for these zones can be far off the "real" behavior of the structure. Note that for such models relationships between the interactions of the high shear forces with flexure and axial ones can be defined.

Using distributed inelasticity models leads to a longer time for computation and to an increased storage data amount. It is considered that the results are more accurate, and there are no prerequisited assumptions or calibrations to be made. Moreover, the "spread" plasticity is considered to be closer to the real behavior of the R/C members subjected to cyclic loadings. On the other hand, the effect of the high stress shear forces cannot be taken into consideration.

Numerical Examples

Numerical Example Nr. 1: For the singly-reinforced rectangular R/C section shown below (Fig. 2.E.1a), for a given maximum strain ε_c in the compressed upper part of the section compute MR and F, according to the procedure in paragraph 2.3.2.

(i) Input data.

  The rectangular section is illustrated in Figure 2.E.1a). The constitutive laws for concrete and for steel are in Figure 2.E.1b) and c).

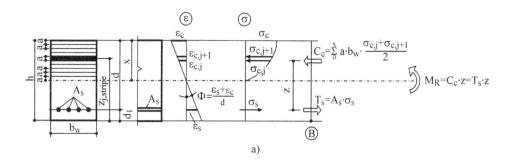

a)

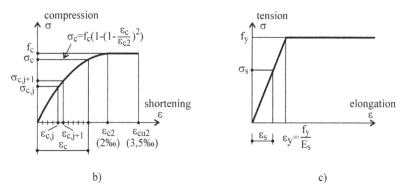

b)   c)

Figure 2.E.1 a) Rectangular single-reinforced element. b) Constitutive law for concrete; c) constitutive law for steel

Geometric data:

$b_w = 250$ mm.
$h = 500$ mm
$d_1 = 50$ mm ($d = h - d_1 = 450$ mm)
$A_s = 1000$ mm^2

Concrete:

$f_c = 25$ MPa (N/mm^2)
$\varepsilon_{c2} = 0.002$
$\varepsilon_{c2u} = 0.0035$

Simplified approach: concrete is working only in compression (tension is neglected).

$\sigma-\varepsilon$ relationship is the parabola-rectangle.

Steel:

$$f_y = 300\,\text{MPa}$$
$$E_s = 200\,\text{GPa}$$

Maximum concrete strain ε_c ranges for 0.00035–0.0035 stepped in 0.00035 (1/10 ε_{c2u}).

In the following, a "by hand procedure" using a spreadsheet program is shown (Fig. 2.E.2). A computer program for the described procedure would lead to more accurate results.

(ii) procedure:

1. For a given ε_c value,
2. Choose a x value, x ranging from d = 450 mm to 22.5 mm (step d/20)
3. We split the depth of the compressed zone x in equal stripes, having the width equal to b_w and the height a = x/10.
4. For each stripe we compute the strain values $\varepsilon_{c,j}$. Strain distribution is linear, ranging from 0 to e_c in (1), step $\varepsilon_c/10$.
5. Compute the stress values $\sigma_{c,j}$ top and bottom of the stripe, by using the $\sigma-\varepsilon$ relationship in Figure 2.E.1b).
6. For each stripe compute the compressive force in concrete as being the area of the stripe times the average stress in the stripe:

$$C_{c,\text{stripe}} = ab_w(\sigma_{c,j} + \sigma_{c,j+1})/2$$

7. By summation, determine the total compressive force in concrete:

$$C_c = \Sigma C_{c,\text{stripe}}$$

8. Compute strain in the reinforcement layer (Bernoulli's assumption):

$$\varepsilon_s = \varepsilon_c(d - x)/x$$

9. By using the $\sigma-\varepsilon$ relationship for steel in Figure 2.E.1c) determine the stress value σ_s ($e_y = f_y/E_s$ – Hook's relationship)
10. Compute the tensile force in steel – steel area times stress level:

$$F_s = A_s\sigma_s$$

11. If x is correctly chosen, compression in concrete must be equal to tension in steel. We check:

$$C_c - T_s < \text{chosen tolerance}$$

12. If NO, back to step (2) for a new x value.

(iii) final:

13. If YES, interpolate for correct x, ε_s and σ_s.
14. Repeat steps (3)–(10) except step (7) – split x in 10 stripes, compute $\varepsilon_{c,j}$, then $\sigma_{c,j}$, compressive force in stripe $C_{c,\text{stripe}}$, ε_s, σ_s and tensile force T_s.
15. Compute M_R, for example in point B (compression and tension are rotating opposite, this is the minus summation below):

$$M_R = \Sigma C_{c,\text{stripe}}\ z_{j,\text{stripe}} - T\sigma d_1 \quad \text{(see Fig. 2.E.1a)}$$

(i) INPUT DATA

GEOMETRY			CONCRETE			STEEL		
bw =	250	mm	fcd (N/mm2) =	25		fy (N/mm2) =	300	
h =	500	mm	ε_{c2} =	0,002		Es =	200000	
d =	450	mm	ε_{cu2} =	0,0035		epsys (fy/Es) =	0,0015	
As =	1000	mm2	$\varepsilon_{c,given}$ =	0,00175		Ec (N/mm2) =	31000	
step x =	22,5	mm						

STEP (1): 0,00175

(ii) PROCEDURE

STEP (4) compute ε_{cj} / STEP (5) compute σ_{cj}

col	ε_{c1}	ε_{c2}	ε_{c3}	ε_{c4}	ε_{c5}	ε_{c6}	ε_{c7}	ε_{c8}	ε_{c9}	ε_{c10}	ε_{c11}
ε	0	0,000175	0,00035	0,000525	0,0007	0,000875	0,00105	0,001225	0,0014	0,001575	0,00175
σ (N/mm2)	0,00000	4,18359	7,98438	11,40234	14,43750	17,08984	19,35938	21,24609	22,75000	23,87109	24,60938

(The σ_{c1}–σ_{c11} values above are constant for all rows 1.0–20.0.)

STEP (2) choose x — STEPS (8) and (9) compare — STEP (7) Cc — STEP (10) Fs — STEP (11) Cc–Ts

step	x (mm)	ε_s (—)	σ_s (N/mm2)	Cc COMPRESSION (N)	Fs TENSION (N)	Cc–Ts (N)
1.0	450,0	0,00000	0,00	1739575,20	0,00	1739575,20
2.0	427,5	0,00009	18,42	1652596,44	18421,05	1634175,38
3.0	405,0	0,00019	38,89	1565617,68	38888,89	1526728,79
4.0	382,5	0,00031	61,76	1478638,92	61764,71	1416874,21
5.0	360,0	0,00044	87,50	1391660,16	87500,00	1304160,16
6.0	337,5	0,00058	116,67	1304681,40	116666,67	1188014,73
7.0	315,0	0,00075	150,00	1217702,64	150000,00	1067702,64
8.0	292,5	0,00094	188,46	1130723,88	188461,54	942262,34
9.0	270,0	0,00117	233,33	1043745,12	233333,33	810411,78
10.0	247,5	0,00143	286,36	956766,36	286363,64	670402,72
11.0	225,0	0,00175	300,00	869787,60	300000,00	569787,60
12.0	202,5	0,00214	300,00	782808,84	300000,00	482808,84
13.0	180,0	0,00263	300,00	695830,08	300000,00	395830,08
14.0	157,5	0,00325	300,00	608851,32	300000,00	308851,32
15.0	135,0	0,00408	300,00	521872,56	300000,00	221872,56
16.0	112,5	0,00525	300,00	434893,80	300000,00	134893,80
17.0	90,0	0,00700	300,00	347915,04	300000,00	47915,04
18.0	67,5	0,00992	300,00	260936,28	300000,00	-39063,72
19.0	45,0	0,01575	300,00	173957,52	300000,00	-126042,48
20.0	22,5	0,03325	300,00	86978,76	300000,00	-213021,24

(iii) FINAL

STEP (13) interpolate for x: x = 77,60515347 mm

Interpolate values:

effort σ_c (N/mm2)	0,00000	4,18359	7,98438	11,40234	14,43750	17,08984	19,35938	21,24609	22,75000	23,87109	24,60938
stripe NUMBER		1	2	3	4	5	6	7	8	9	10
lever arm zj (mm)		426,28	434,04	441,80	449,56	457,32	465,08	472,84	480,60	488,36	496,12
Cc;stripe (N)		4058,36	11803,71	18806,37	25066,31	30583,55	35358,09	39389,92	42679,05	45225,46	47029,18
Cc;stripe zj; stripe (Nmm)		1729975,887	5123232,114	8308579,831	11268727,71	13986384,4	16444258,59	18625058,93	20511494,09	22086272,73	23332103,53

$\sum$ Cc;stripe · zj; stripe (Nmm) = 141416088

Ts · d1 (Nmm) = 150000000

STEP (14)

STEP (15): MR = 126,4160878 KNm

STEP (16): φ = 0,0000226

STEP (13) interpolate for ε_s and σ_s:

ε_s	σ_s
0,008398	300

Figure 2.E.2 Example spreadsheet for $\varepsilon_c = 0.00175$

Table 2.E.1 M_R and Φ computation for different values of ε_c

Step	ε_c	X (mm)	ε_s	C_c (KN)	T_s (KN)	M_R (KNm)	Φ (1E-5)
1	0.00035	144.69	0.00074	148.97	147.71	59.77	0.242
2	0.00070	143.60	0.00146	281.15	292.69	111.79	0.480
3	0.00105	110.94	0.00321	300.00	300.00	123.27	0.946
4	0.00140	89.58	0.00563	300.00	300.00	125.33	1.563
5	0.00175	77.61	0.00840	300.00	300.00	126.42	2.255
6	0.00210	70.51	0.01130	300.00	300.00	126.98	2.978
7	0.00245	66.13	0.1422	300.00	300.00	127.26	3.705
8	0.00280	63.19	0.1714	300.00	300.00	127.42	4.431
9	0.00315	61.09	0.02005	300.00	300.00	127.51	5.157
10	0.00350	59.51	0.02297	300.00	300.00	127.57	5.881

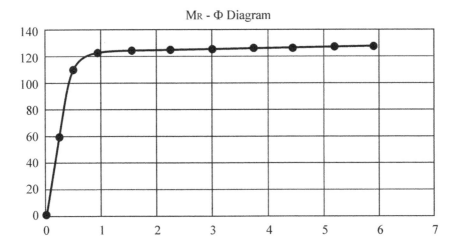

Figure 2.E.3 $M_R - \Phi$ graph

16. Compute the curvature Φ for the correct value of x:

$$\Phi = (\varepsilon_s + \varepsilon_c)/d$$

For strain ε_c ranging from 0.00035 to 0.0035 stepped in 0.00035, the results are shown in Table 2.E.1 and Figure 2.E.3.

Numerical Example Nr. 2: For the rectangular R/C section shown below (Fig 2.E.4a), for a given axial force varying from N = 0–3900 KN step 300 KN, compute M_R and Φ and trace N – M_R and N – Φ diagrams, according to the procedure in paragraph 2.3.3.

(i)   Input data.
       The rectangular section is illustrated in Figure 2.E.4a. The constitutive law for concrete is in Figure 2.E.4b and for steel in Figure 2.E.4c.

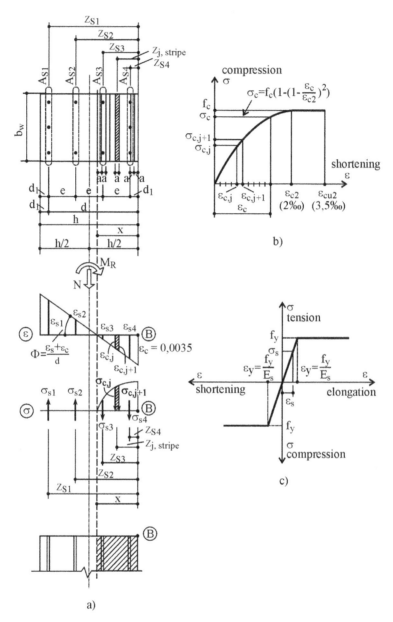

a)

Figure 2.E.4 a) Rectangular R/C element. b) Constitutive law for concrete. c) Constitutive law for steel

Geometric data:
$b_w = 400$ mm.
$h = 600$ mm
$d_1 = 50$ mm ($d = h - d_1 = 550$ mm)
$e = 500/3 = 166.6$ mm
$A_{s1} = 900$ mm^2; $A_{s2} = 600$ mm^2; $A_{s3} = 600$ mm^2; $A_{s4} = 900$ mm^2

Concrete:

$f_c = 20\,\text{MPa}\ (\text{N/mm}^2)$

$\varepsilon_{c2} = 0.002$

$\varepsilon_{c2u} = 0.0035$

Simplified approach: concrete is working only in compression (tension is neglected).

$\sigma-\varepsilon$ relationship is the parabola-rectangle.

Steel:

$f_y = 300\,\text{MPa}$

$E_s = 200\,\text{GPa}$

Simplified approach: steel is working similar both in tension and in compression.

$\sigma-\varepsilon$ relationship is the elastic-perfect plastic relationship.

$\varepsilon_y = f_y/E_s$ – Hook's relationship.

General Assumption:

M_R is reached in the moment the maximum strain $\varepsilon_{c2u} = 0.0035$ is reached in the most compressed fiber of the section.

In the following, a "by hand procedure" using a spreadsheet program is shown (Fig. 2.E.5).

(ii)   Procedure:

1. For a given N value (varying, from 0 to 3900 KN step 300 KN),
2. Choose a x value, x ranging from $d = 550\,\text{mm}$ to $27.5\,\text{mm}$ (step d/20)
3. Split the depth of the compressed zone x in equal stripes, having the width equal to σ_w and the height $a = x/10$.
4. For each stripe compute the strain values $\varepsilon_{c,j}$. Strain distribution is linear, ranging from 0 to $\varepsilon_c = \varepsilon_{c2u} = 0.0035$.
5. Compute the stress values $\sigma_{c,j}$ left and right of the stripe, by using the $\sigma-\varepsilon$ relationship in Fig. 2.E.4b.
6. For each stripe compute the compressive force in concrete as being the area of the stripe times the average stress in the stripe:

$C_{c,stripe} = ab_w(\sigma_{c,j} + \sigma_{c,j+1})/2$

7. By summation, determine the total compressive force in concrete:

$C_c = \Sigma C_{c,stripe}$

8. Compute strain in the reinforcement layers (Bernoulli's assumption):

$\varepsilon_{s1} = \varepsilon_{c2u}(z_{s1} - x)/x;\ \varepsilon_{s2} = \varepsilon_{c2u}(z_{s2} - x)/x;\ \text{etc.}$

9. By using the $\sigma-\varepsilon$ relationship for steel in Fig. 2.E.4c we determine the stress values $\sigma_{s1};\ \sigma_{s2};$ etc., paying attention to the sign (compression or tension),
10. Compute the total force in steel – steel area times stress level:

$F_s = \Sigma A_{s,i}\sigma_{s,i}$ (attention:compression or tension)

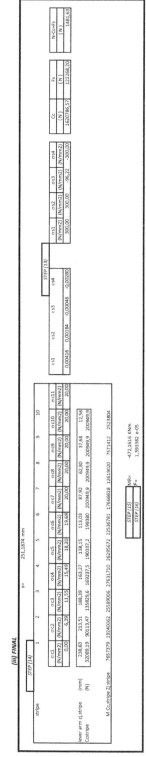

Figure 2.E.5 Example spreadsheet for N = 1500 kN.

Table 2.E.2 N, M_R and F computation for different values of N

Step	N (KN)	x (mm)	M_R (KNm)	Φ (1e-05)
1	0	71.26	233.47	4.9114
2	300	103.13	304.46	3.3938
3	600	146.36	365.51	2.3914
4	900	180.76	404.39	1.9363
5	1200	214.66	441.18	1.6305
6	1500	251.19	472.16	1.3934
7	1800	285.75	488.85	1.2248
8	2100	318.23	497.12	1.0998
9	2400	352.62	500.23	0.9926
10	2700	388.52	494.26	0.9009
11	3000	419.92	467.57	0.8335
12	3300	452.82	436.88	0.7729
13	3600	487.05	401.37	0.7186
14	3900	522.48	360.24	0.6699

11. If x is correct chosen, the sum of the internal forces must be equal to the external given axial force N in step (1). We check:

$$N - C_c + F_s < \text{chosen tolerance.}$$

12. If NO, back to step (2) for a new x value.

(iii)   final:

13. If YES, interpolate neighborhoods for correct $x, \varepsilon_{s1}...\varepsilon_{s4}$ and $\sigma_{s1}...\sigma_{s4}$
14. Repeat steps (3) – (10) except step (7) – split x in 10 stripes, compute $\varepsilon_{c,j}$, then $\sigma_{c,j}$, compressive force in stripe $C_{c,\text{stripe}}$, $\varepsilon_{s1}...\varepsilon_{s4}$ and $\sigma_{s1}...\sigma_{s4}$ and steel forces $A_{s,1}\sigma_{s,1} ... A_{s,4}\sigma_{s,4}$ (attention: compression or tension).
15. Compute M_R, for example in point B (paying attention to the direction for the forces):

$$M_R = \Sigma C_{c,\text{stripe}}\, z_{j,\text{stripe}} + \Sigma A_{s,i}\sigma_{s,i} z_{s,i} + N(h/2 - d_1) \text{ (see Fig. 2.E.4a))}$$

16. Compute the curvature Φ for the correct value of x:

$$\Phi = (\varepsilon_{s1} + \varepsilon_{c2u})/d$$

For strain N ranging from 0 to 3900 KN stepped in 300 KN, the results are shown in Table 2.E.2 and Figure 2.E.6.

Numerical Example Nr. 3: For the cantilever column shown below (Fig. 2.E.7a), assuming the same cross section and material characteristics as in Numerical Example 2, compute the yielding initialization moment M_y and yielding curvature Φ_y, then the ultimate (resistance) values M_u and Φ_u. Assuming the height of the column $l = 5.00$ m and the length of the plastic zone $l_p = h = 600$ mm, compute the correspondent top-displacements Δ_y and Δ_u, according to the procedure in paragraph 2.4.3. The computations are made for two variants, firstly considering $N = 0$ (pure bending) and

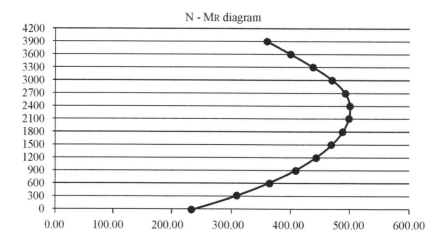

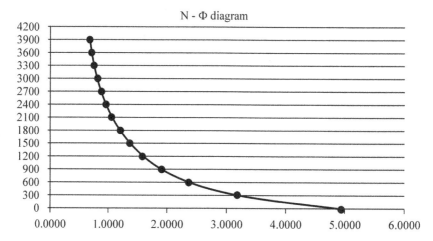

Figure 2.E.6 a) N – M_R and b) M_R – Φ diagrams

second for N = 1500 kN (eccentrically compression). Compute the ductility factors μ_Φ and μ_Δ (curvature and displacements) and trace the M – Δ diagrams.

(i)   Procedure:

The flow of the computations for finding the depth of the compressed concrete zone x, corresponding bending moment value M and curvature Φ is the same to that shown in the Numerical Example 2.

The difference consists that for finding the yielding initialization values we will use the strain diagram in Figure 2.E.8a, considering $\varepsilon_{s1} = \varepsilon_y$ and ε_c unknown, and for the ultimate ones we use the strain diagram in Figure 2.E.8b, considering known $\varepsilon_{c2,u} = 0.0035$ and the values in steel $\varepsilon_{s1\ldots4}$ unknown, but in respect to Bernoulli's assumption.

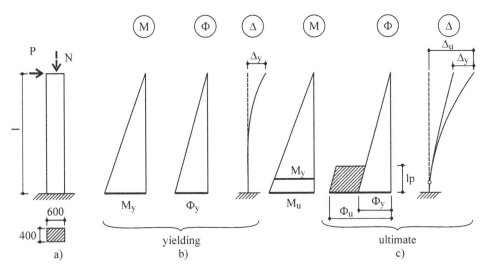

Figure 2.E.7 a) cantilever column; b) yielding initializaton diagrams; c) ultimate (resistance) diagrams

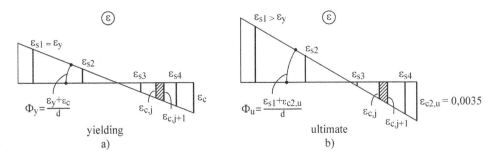

Figure 2.E.8 Strain diagrams for: a) yielding initialization; b) ultimate

The spreadsheet used for the "by hand procedure" is the same as for Example 2.

Computations are made for both axial force values ($N = 0$ and $N = 1500\,KN$)

The final results are presented in the table and diagrams below.

(ii) Finding the top-displacements Δ:

The displacements are computed by integration of the curvature diagrams along the height l of the cantilever:

– the yielding value $\Delta_y = \Phi_y\, l^2/3$

– the ultimate value $\Delta_u = \Delta_y + (\Phi_u - \Phi_y)l_p\,(1 - l_p/2)$

(iii) Final:

We compute the ductility factors:

– curvature ductility factor $\mu_\Phi = \Phi_u/\Phi_y$

– displacement ductility factor $\mu_\Delta = \Delta_u/\Delta_y$

Table 2.E.3 M, Φ and Δ values for yielding initialization and ultimate

N (KN)	M_y (KNm)	Φ_y (1e-05)	Δ_y (mm)	M_u (KNm)	Φ_u (1e-05)	Δ_u (mm)
0	173.19	0.3666	30.55	233.47	4.9114	158.71
1500	429.02	0.5833	48.61	472.16	1.3934	76.31

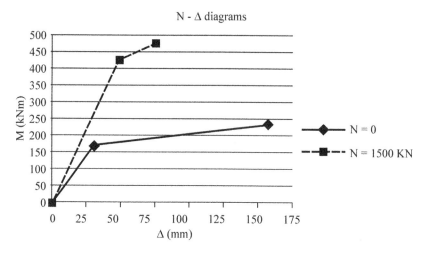

Figure 2.E.9 M–Δ diagrams for N = 0 and N = 1500 KN

Results:

for N = 0, $\mu_\Phi = 13.39$ and $\mu_\Delta = 5.19$
for N = 1500 KN, $\mu_\Phi = 2.39$ and $\mu_\Delta = 1.57$

One can observe the ductility capacity reduction of the element due to the large axial force magnitude.

Conclusions

Explosive developments during the last decades in computer science provided powerful tools for reinforced concrete behavior modeling. Instead of the old assumption of pure elastic behavior, advanced non-linear structural models have been developed for reinforced concrete based on accurate quantification of the effect of numerous significant parameters. The purpose of the present chapter is to present an overview of the up-to-date models and methods which allow understanding and assessing accurately the response of reinforced concrete under different types of loadings. Theoretical considerations are illustrated through some numerical examples.

Chapter 3

Behavior and Analysis of Reinforced Concrete Structures under Static Loads

Abstract

Based on the reinforced concrete constitutive laws presented in the previous chapter of the book, behavior of R/C structures can be foreseen. So, it was stated that the fundamental feature of reinforced concrete structure behavior is its non-linear character. Within the present chapter methods for elastic and post-elastic analysis under static loads are examined as applied to reinforced concrete frames. Initially, "classical" push-over analysis is examined. Further, the plastic analysis through "limit equilibrium" (sometimes referred to as Simple Plastic Theory) with its theorems and practical applications is briefly presented and commented. It is demonstrated that the rotations of plastic hinges depends upon the moments plastic redistribution only and, accordingly, practical procedure for post-elastic design of concrete structures which keep under control the magnitude of plastic rotations is derived. Some numerical examples aimed to illustrate the theoretical considerations are included.

3.1 Behavior of Reinforced Concrete Structures under Monotonic Loads

The constitutive laws of reinforced concrete *sections* and *elements* are the background for understanding and modeling the specific behavior of different types of *structures*, made of this material, loaded progressively up to the failure. For starters let us focus our attention to the structures loaded *monotonically* and *statically*.

Consider a simple reinforced concrete structure – a portal frame – loaded as shown in Figure 3.1. It is supposed that geometrical dimensions and the reinforcement of the elements (beam and columns) are known. We accept that the external loads increase progressively varying proportionally to each other up to the failure (collapse) of the structure:

$$S = \lambda S_0 \tag{3.1}$$

$$P = \lambda P_0 \tag{3.2}$$

where: λ is a proportionality coefficient, called *loading factor*, with magnitude which varies from zero up to the value corresponding to the failure of the structure: $\lambda = 0, \ldots, \lambda_u$, S_0 and P_0 – initial magnitude of forces S and P.

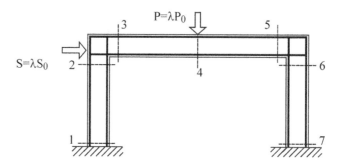

Figure 3.1 Example of reinforced concrete frame

According to the magnitude of incremental external load (respectively of the loading factor λ) the internal forces (M, N, and V) will increase in each section of the structure. Accordingly, the structure sections behave within the stages described by the *sectional constitutive laws* (see Chapter 2).

The overall behavior of the structure results from the behavior of each section. Accepting members' ideal elastic-plastic model, when the external loads increase, the structure will be in one of the following behavior stages:

1. *Elastic stage.* All structure's members behave elastically. If the structure is unloaded, in this stage, no residual deformations are recorded. This behavior stage characterizes the behavior of the structures under service loads (gravity and wind loads).
2. *Elastic-plastic stage.* If the external loads further increase, at a certain value of the loading factor: $\lambda = \lambda_p$, in a section of the structure the plastic moment M_p is reached. In this section the flexure reinforcement yields and the sectional rigidity drops suddenly even though the structure is able to carry its external loads. If the external loads further increase, the plastic deformations are propagated along *the plastic length l_p.* The behavior of the element is that described in Chapter 2, referring to *constitutive laws of the elements.*

   Quantitatively speaking, the behavior of the element, which has a plastic zone, can be described through the *post-elastic stiffness matrix of the element* (see Chapter 2). The matrix *stiffness* of the *structure* will be accordingly modified.

   A suggestive representation of the structure behavior, after occurrence of plastic deformations, is obtained by modeling the plastic zones by *plastic hinges* (see Chapter 2) (Fig. 3.2). According to this assumption, in each critical section where moment capacity is reached plastic hinge occurs, i.e. hinges acted by a constant moment which is the plastic moment M_p. Consequently, in these sections, the structure looses the full continuity and the degree of static indeterminacy is decreased. The structure becomes more flexible but it can still resist external loads.

   If the external loads further increase, plastic hinges successively appear in other sections. When the number of plastic hinges is equal to the degree of static indeterminacy, the structure becomes *static determinate.*

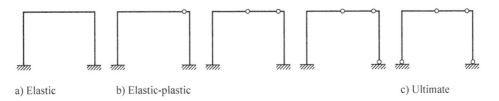

a) Elastic   b) Elastic-plastic   c) Ultimate

Figure 3.2 Progressive development of plastic hinges

During the gradual increase of external loads, within plastic hinges *plastic rotations* are developed. The occurrence of large number of plastic hinges is possible provided that their *rotation capacity* is not exceeded. When this condition is no more fulfilled, the zone in which plastic deformation is bigger than the capable one *fails* and the element is out of work. Another condition to be observed is *the avoidance of premature brittle failure of certain elements of the structure.* This could occur when flexural reinforcement percentage is too small or too big, when the element has not enough shear capacity, for elements with too high level of compressive axial force, or when buckling is not properly prevented, for elements with inadequate constructive details, etc. Failure of a vital element leads to *the failure of the structure* (for example, the case of failure of a frame column). Nevertheless, there are situations when the structure *survives* even after failure of an element provided that this element is not vital for the overall resistance and stability of the structure.

3. *Ultimate stage.* When the external incremental loading generates a new plastic hinge within the static determinate structure, this is transformed into a mechanism which can no longer resist external loads. This is equivalent to total collapse. We say that the structure has reached its *ultimate stage.*

The ultimate stage is reached by the transformation of the structure into a mechanism provided that premature rupture doesn't occur. Ultimate stage can be reached prematurely too if a local mechanism is developed (Fig. 3.3).

3.2 Methods for Static Analysis of Reinforced Concrete Structures

Different types of methods for structural analysis of reinforced concrete correspond to different behavior stages above defined. In other words, each design and analysis approach accepts a certain behavior stage as reference stage for the structural analysis.

The adopted reference stage implies consequences on the "safety coefficients" system, which quantifies the global reliability of the structure regarding the adopted reference stage. When selecting one or another type of method of structural analysis, the designer should be aware that he should use magnitude of external loads, sectional stiffness, strength of materials, etc., consistent to the safety for that specific stage. If this requirement is ignored, we can easily reach over- or under-evaluated safety.

Another aspect, which should be taken into account when selecting the method for static analysis is that for reinforced concrete structures interdependence exists between

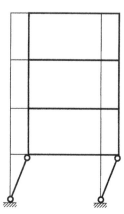

Figure 3.3 Premature failure through local mechanism

internal forces, proportion and detailing of structural components (geometrical dimensions, reinforcement percentages) and the loading stage. Depending upon the nature of input data *initially known* and the values to be determined through the structural analysis, two types of problems are encountered: *structural design* and *verification*.

Structural design is specific to new constructions. In these situations *external loads* are known and *internal forces* at critical sections of the structure are required. They allow proportioning of the concrete sections and to determine the reinforcement amount and detailing.

Verification is referring to *existent* structures; the analysis is performed in order to evaluate if the structure is complying with strength and deformability requirements. In this case, the effective sectional capacities of all critical sections are given (determined by taking into account the existent concrete sections and reinforcement amount) and the magnitude of external loads able to be carried out by the structure ("*structure capacity*" or "*resistance*") is required.

Specific methods of static analysis have been developed for both of these two problems.

3.2.1   Static Elastic Analysis

If elastic behavior stage is accepted as reference stage, then the current methods of static (elastic) analysis can be used for reinforced concrete structures too. However, some adjustments have to be implemented, which quantifies the behavior peculiarities of the reinforced concrete sections under service loads as well as specific modeling of different types of structures (cast-in-place or prefabricated frames, systems with structural walls, etc.). These aspects will be analyzed in detail in the following chapters.

Under service loads it is accepted that the reinforced concrete structures behave linearly elastic, even though there is small physical non-linearity because of the cracking or other similar phenomena. Accordingly, for the analysis under service loads, elastic linear structural analysis can be accurately used.

3.2.2 Static Post-Elastic ("Pushover") Analysis

The proper structural analysis within elastic-plastic domain of behavior *without reaching the ultimate stage* should take into account explicitly the progressive development of plastic deformations. This can be done through *static post-elastic analysis*. Such analysis quantifies the resistance reserves of an existing structure or of a structure already dimensioned. Thus, the static post-elastic analysis refers to the *verification* problem.

Comments

Difference should be noted between *elastic, post-elastic* and *non-linear* behavior and analysis.

Elastic behavior is that for which no residual deformations exist when unloading the structure.

Non-linear analysis refers to the non linear relationships between forces and displacements. The non linearity can have *physical* or *geometrical* causes.

The *physical non-linearity* comes from the non-linearity of the *constitutive laws*. It can be *elastic non-linearity*, when loading and un-loading follow the same (non-linear) path or *post-elastic non-linearity* (with residual deformations after total unloading). Substantial post-elastic non-linearity appears in reinforced concrete sections after yielding of the flexural reinforcement.

The *geometrical non-linearity* is related to the occurrence of large deformations or modifications in the structure geometry (foundation detachment from soil, buckling, etc). In these cases, the equilibrium of the element or of the structure must be written on the deformed position ("second order analysis"). The geometrical non linearity is specific, for example, to slender elements eccentrically compressed.

Static post-elastic analysis is specific to structures loaded with variable actions (live, wind, snow forces) acting together with permanent ones.

This type of analysis has a significant importance for structural systems subjected to seismic actions of high intensity, for which the appearance of substantial post-elastic deformations is accepted through the general design philosophy itself. For these structures specific advanced procedures have been developed. They will be examined within Chapter 4.

The static post-elastic analysis is performed through "step-by-step" procedures ("biographic" or "pushover" approach) that follow the progressive development of plastic deformations.

The main idea is to determine – step-by-step – the internal forces and deformations of the structure progressively loaded, following the change of its global rigidity, due to successive occurrence of plastic deformations.

The loading that varies monotonically, generating gradual plastic deformations, can be a system of *forces* or of *imposed displacements*. Besides the variable loads, a system of constant loads can be considered.

For illustrating the static post-elastic analysis, we consider a structure (frame) subjected to permanent constant loads $\{G\}$ and horizontal forces $\{H\}$ which increase proportionally with a parameter λ called *loading factor*: $\{H\} = \lambda\{H_0\}$ (where $\{H_0\}$ is the initial value of the horizontal loads vector) (Fig. 3.4).

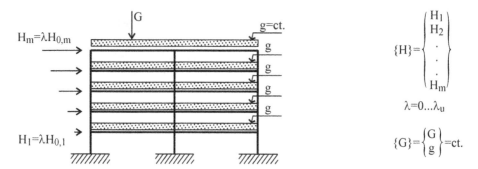

Figure 3.4 Frame subjected to permanent constant loads and horizontal forces which increase proportionally with a parameter

It is accepted that the elements of the structure are subjected predominantly to bending. The bending moment can be accompanied or not by axial forces (compression or tension).

Given:

- the structure (geometry and topology)
- concrete sections and the reinforcement areas of all critical sections.

Accordingly, the flexural capacities of critical sections (capable moments), with their signs, positive or negative, depending on the fiber in tension, are known:

$$\{M_R\} = M_{R,i} \tag{3.3}$$

- the system of gravity loads (constants) $\{G\}$ and the initial magnitudes of variable loads $\{H_0\}$

Required:

- the magnitude of the variable loads $\{H_u\} = \lambda_u\{H_0\}$, respectively of the loading factor λ, at each loading step for which a new plastic hinge occurs until the ultimate stage (collapse) of the structure is reached.

According to the way it was defined, this method is a typical *verification* approach. It leads to determination of the structure resistance.

The process (methodology), which will be presented below, accepts as basic assumption that the plastic deformations are concentrated in *plastic hinges*, located in some critical sections. When the bending moment in these sections reaches its capacity M_R, plastic hinge occurs.

The calculation steps are the following:

(i) Static analysis under gravity loads is performed, determining the internal forces at critical sections, i.e. the bending moments $\{M_G\}$;

(ii) Static analysis is performed under initial magnitudes of variable loads $\{H_0\}$, resulting the bending moments $\{M_H\}$;

(iii) Capable moments at all critical sections are determined, as function of section characteristics (geometry, reinforcement amount) and of the magnitude of the axial force $N_{G,i} + N_{H,i}$

Note: In each section the capable moments have to be considered according to the two possible senses of the action, positive and negative. Corresponding to the reinforcement areas on the two sides of the section, the capable moment has usually two values. To the two values are attached the sign plus or minus.

(iv) The ratios γ_i are computed

$$\gamma_i = \frac{M_{H,i}}{M_{R_i} - M_{g,i}} \quad (i = 1, \ldots, n) \tag{3.4}$$

Note: The capable moment M_R is that which has the same sign as the moment generated by the variable loads H.

Explanation

The coefficients γ_i represent the ratios between the moments generated by the variable loads and the "reserve of strength" of that section. At each section "reserve of strength" consists of the capable moment $M_{R,i}$ to which is subtracted or added the moment from the permanent load $M_{G,i}$ as it has opposite sign from the $M_{R,i}$. By comparing the reserve of strength $(M_{R,i} - M_{G,i})$ with $M_{H,i}$ the section which yields first is determined as being the one for which the ratio γ_i has highest magnitude.

(v) The maximum magnitude of the ratio γ_i is selected:

$$\max(\gamma_i) = \gamma_k \tag{3.5}$$

In section k a plastic hinge will occur.

(vi) The magnitude of the load $H^{(1)}$, that produces the plastic hinge, is determined:

$$H^{(1)} = \frac{1}{\gamma_k} \cdot \{H_0\} = \lambda^{(1)}\{H_0\} \qquad \lambda^{(1)} = \frac{1}{\gamma_k} \tag{3.6}$$

(vii) A plastic hinge is introduced in section "k". Accordingly, the stiffness matrix of the element containing the plastic hinge is modified and, consequently, a new stiffness matrix of the structure is generated as well.

(viii) Go to step (ii) (with new stiffness matrix of the structure).

The cycle (ii)–(viii) is repeated until a number of plastic hinges that transforms the structure into a mechanism is reached. The stiffness matrix of the structure transformed into a mechanism becomes singular (it cannot be inverted anymore).

Actually, before step (viii), supplementary steps have to be performed in order to check-up following conditions:

– if the rotations in the plastic hinges are below their rotational capacity;
– the shear forces in the critical sections do not overcome the resistance.

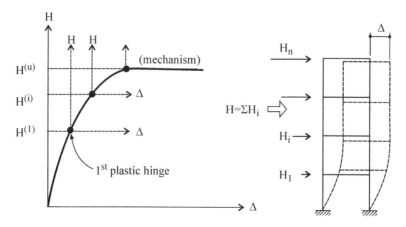

Figure 3.5 Total variable force/displacement relationship

If these conditions are not fulfilled, premature failure of the element is expected and, accordingly, it has to be decided whether:

- the analysis is continued, eliminating from the structure the failed element, or
- the analysis is stopped, considering that the ultimate stage of the structure has been reached.

A similar approach has been implemented in many computer analysis programs.

Nowadays advanced software is available which accepts loading with incremental variable *displacements* or elements with *distributed plasticity* (see Chapter 2).

The overall quantitative response of the structure to the variable external loads $\{H\}$ can be visualized through the $H-\Delta$ diagram where: $H = \Sigma H_i$ is the total variable loading on the structure and Δ is the horizontal displacement at the structure top.

It can be remarked that the static post-elastic analysis, under the assumption of plastic hinges, is equivalent to successive elastic analysis of the structure, having hinges located in the sections that successively yield (Fig. 3.5).

3.2.3 *Post-Elastic Analysis through "Limit Equilibrium"*

The *ultimate stage* of a structure can be *directly* quantified, without explicit analysis of intermediary stages that lead the structure from elastic to the ultimate stage.

The theoretical background of the method is the *Theory of Limit Analysis*, a component of *Plastic Analysis of Structures*. Sometimes it is referred to as *Simple Plastic Theory* (Hodge, 1981).

That theory applies to structures made by ideal elastic-plastic materials, subjected to loads that increase proportionally with a single parameter, at which the plastic zones are modeled with point-plastic hinges.

For such structures it can be demonstrated that, if: (a) external loads are in equilibrium with the internal forces ("static" condition), (b) in no sections of the structure the plastic moment is exceeded ("yielding" condition) and (c) enough plastic hinges have been developed to transform the structure into a mechanism ("kinematical" condition), then the magnitude of the external loads corresponding to that stage (which is the *ultimate stage*) is *unique* and defined as *ultimate* (failure) load. This statement is referred to as *theorem of the uniqueness of ultimate load*.

It can be also demonstrated that:

- if only the conditions (a) (equilibrium) and (b) (yielding condition) are fulfilled, then the external load is smaller or equal to the ultimate load ("*static theorem*" or "*theorem of lower bound of ultimate load*"),
- if only the conditions (a) and (c) (mechanism) are fulfilled, then the external load is greater or equal to the ultimate load ("*kinematical theorem*" or "*theorem of upper bound of ultimate load*").

Based on each of these theorems procedures for direct determination of the ultimate load can be defined.

The practical procedures for determining the ultimate loads can be classified as follows:

1. For the simple structures, at which the failure mechanism (i.e. location of the plastic hinges) is evident, the ultimate load is determined by writing the equation of equilibrium between the external loads and the plastic moments through virtual work principle. This type of procedures is based on the uniqueness theorem since they consider explicitly all three conditions of this theorem.
2. When the failure mechanism is not evident but the potential positions of the plastic hinges can be appreciated, more possible failure mechanisms are chosen and the condition of equilibrium between the external loads and the plastic moments is written for each chosen mechanism (using virtual work principle). The external load thus determined, is greater or equal to the real ultimate load (unique), according to kinematical theorem.
3. If we choose (arbitrary) bending moments diagrams in equilibrium with the external loads, having magnitudes less or equal to the plastic moments, then, according to the static theorem, the external load is smaller or equal to the ultimate load.

Among these types of procedures, those effectively used for the reinforced concrete structures will be described more detailed below.

Limit Analysis for Simple Structures

For simple structures, as that in Figure 3.6, the location of plastic hinges which transform the structure into a mechanism can be established from very beginning of the analysis. In each plastic hinge acts the plastic moment M_p, supposed to be known. *The sign* of the plastic moment is chosen so that, at a virtual displacement that produces positive mechanical work with the external loads, the virtual work of all positive moments *be negative* (*opposes* to the work of external loads).

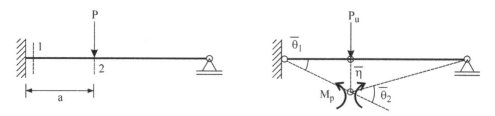

Figure 3.6 Simple structure with known location of plastic hinges

The equilibrium condition between the external loads and the plastic moments can be written using the theorem of virtual work:

$$\sum P_u \cdot \bar{\eta} = \sum M_p \cdot \bar{\theta} \qquad (3.7)$$

Because the external loads vary proportionally with the parameter λ (the *loading factor*), results:

$$\lambda_u \sum P_0 \cdot \bar{\eta} = \sum M_p \cdot \bar{\theta} \qquad \lambda_u = \frac{\sum P_0 \cdot \bar{\eta}}{\sum M_p \cdot \bar{\theta}} \qquad (3.8)$$

where: P_0 is the initial (reference) magnitude of load,
   M_p plastic moments, known from the initial input data;
   $\bar{\eta}$ and θ – virtual displacement and rotation;
   λ_u – the ultimate value of the loading factor.

In this format the method is a *verification* one since it refers to structures *already dimensioned* for which all geometrical and mechanical characteristics are known and the ultimate load magnitude is required.

The method implies the following steps:

(i)   Plastic hinges are considered at appropriate locations (supposed to be known) so that the structure is transformed into a mechanism.

(ii)   Plastic moments M_p are applied in the plastic hinges, having signs that oppose to the external displacement.

(iii)   A virtual displacement is imposed to the mechanism created by introducing plastic hinges. Virtual displacements on direction of each external force and the virtual rotations in plastic hinges are determined.

(iv)   The equation of virtual work is applied, determining the ultimate magnitude of the loading factor λ_u. Accordingly, the magnitude of external load that transforms the structure into a mechanism is determined

The solution is unique and well determined.

Determination of the Plastic Moments for Simple Structures

For a *new* structure, which is in the process of designing, the problem is to determine the values of the necessary *capable moments* required for balancing external loads, which correspond to the ultimate stage.

The structure is transformed into a mechanism through occurrence of $n+1$ plastic hinges (where n is the degree of statically indeterminacy of the structure). For a supposed simple structure, the sections in which appear the plastic hinges are known. Thus, the problem has $n+1$ unknown quantities that are the plastic moments M_{p1}, $M_{p2}, \ldots, M_{p(n+1)}$. The only available relationship is the equilibrium equation (3.1). Consequently we have a single equation with $(n+1)$ unknowns ("algebraic indeterminacy").

Theoretically, for solving the problem, arbitrary n values of M_p should be chosen or n relationships between them. In fact, for reinforced concrete structures, M_p has to observe specific conditions, which will be discussed in paragraphs 3.2.4 and 3.3.

Limit Analysis – General Case

In the general case the location of plastic hinges that transforms the structure into a mechanism is not *apriori* known. Are known only the potential plastic sections (located at the critical sections $1,\ldots, m$, with $m > n$ where n is the degree of static indeterminacy).

In these situations, the analysis is done by trial-and-error procedures:

(i)   A configuration of $(n+1)$ plastic hinges is chosen which transform the structure into a mechanism;
(ii) The equation (3.1) is applied and a $\lambda_u^{(1)}$ is determined;
(iii) The steps (i) and (ii) are repeated several times, determining a set of values for λ_u;
(iv) The minimum value of λ_u is chosen, resulting $\{S_u\} = \lambda_u\{S_0\}$.

More specific procedures for determining the real mechanism can be used, such as the *method of mechanisms combination*. More sophisticated procedures derive from the Theory of Linear Programming (Simplex algorithm) being, generally, implemented into computer analysis software.

3.2.4 Rotation of Plastic Hinges

In contrast with the case of the ideal elastic-plastic structures, studied by the simple plasticity theory, the reinforced concrete structures have a *limited rotation capacity* of the plastic hinges.

The magnitude of the rotation capacity of the plastic hinges is determined by the constitutive laws of the element (see chapter 2) or they could be appreciated, more roughly, by empirical relationships.

The *effective* rotations in the plastic hinges can be exactly determined only by pushover analysis.

When "limit equilibrium" analysis is used, actual rotations in the plastic hinges cannot be explicitly determined and the risk does exist of premature failure due to excessive plastic rotations. In order to avoid such phenomena, rules that ensure development of limited plastic rotations specific to reinforced concrete structures, have to be defined. For this purpose, the factors that determine the magnitude of the plastic hinge rotation have to be firstly examined.

Let us consider a structure in which, under the external loads that increase proportionally, n plastic hinges have appeared (n is the degree of statically indeterminacy). In this case the structure becomes statically determined.

We also suppose that under progressively increasing loads, all plastic hinges rotate in the same sense (meaning that there are no plastic hinges which close under monotonic loading). This assumption is relatively restrictive (is not evidently respected by any structure, under any type of load), but the conclusions resulting from it remain qualitatively valid for the structures to which they are not accurately fulfilled.

The mutual rotation angle of the sections adjacent to the plastic hinge is called *plastic rotation* and it is denoted by θ_p.

The structure with n plastic hinges behaves as an elastic structure, statically determined, loaded with the external forces and with the n plastic moments acting within the hinges. Applying the principle of effects superposition, the rotation in each plastic hinge can be computed by summation of the effects of the plastic moments and of the external loads:

$$\begin{cases} \theta_{11} \cdot M_{p,1} + \theta_{12} \cdot M_{p,2} + \cdots + \theta_{1n} \cdot M_{p,n} + \theta_{10} = \theta_{p,1} \\ \theta_{21} \cdot M_{p,1} + \theta_{22} \cdot M_{p,2} + \cdots + \theta_{2n} \cdot M_{p,n} + \theta_{20} = \theta_{p,2} \\ \vdots \\ \theta_{n1} \cdot M_{p,1} + \theta_{n2} \cdot M_{p,2} + \cdots + \theta_{nn} \cdot M_{p,n} + \theta_{n0} = \theta_{p,n} \end{cases} \tag{3.9}$$

where: $\theta_{i,j}$ represents the rotation in hinge i generated by the pair of moments $M_j = 1$, applied to the hinge j;

$\theta_{i,0}$ – the rotation in section i due to the external loads.

If, instead of the moments M_p, we assume that the structure is acted by the bending moments M_{el}, corresponding to the *elastic* behavior of the structure, the rotations are equal to zero:

$$\begin{cases} \theta_{11} \cdot M_{el,1} + \theta_{12} \cdot M_{el,2} + \cdots + \theta_{1n} \cdot M_{el,n} + \theta_{10} = 0 \\ \theta_{21} \cdot M_{el,1} + \theta_{22} \cdot M_{el,2} + \cdots + \theta_{2n} \cdot M_{el,n} + \theta_{20} = 0 \\ \vdots \\ \theta_{n1} \cdot M_{el,1} + \theta_{n2} \cdot M_{el,n} + \cdots + \theta_{nn} \cdot M_{el,n} + \theta_{n0} = 0 \end{cases} \tag{3.10}$$

By subtracting the two set of equations results:

$$\begin{cases} \theta_{11} \cdot (M_{p,1} - M_{el,1}) + \theta_{12} \cdot (M_{p,2} - M_{el,2}) + \cdots + \theta_{1n} \cdot (M_{p,n} - M_{el,n}) = \theta_{p,1} \\ \theta_{21} \cdot (M_{p,1} - M_{el,1}) + \theta_{22} \cdot (M_{p,2} - M_{el,2}) + \cdots + \theta_{2n} \cdot (M_{p,n} - M_{el,n}) = \theta_{p,2} \\ \vdots \\ \theta_{n1} \cdot (M_{p,1} - M_{el,1}) + \theta_{2n} \cdot (M_{p,2} - M_{el,2}) + \cdots + \theta_{nn} \cdot (M_{p,n} - M_{el,n}) = \theta_{p,n} \end{cases} \tag{3.11}$$

If we denote by $\overline{M} = M_p - M_{el}$ and we write the set of equations in matrix format, the outcome is:

$$[\theta] \cdot \left\{\overline{M}\right\} = \left\{\theta_p\right\} \tag{3.12}$$

where $[\theta]$ – is *the flexibility matrix* of the structure, corresponding to the statically determined structure obtained by introducing n hinges in the sections in which plastic hinges will appear.

From the equations (3.5) or (3.6) results that the rotations in the plastic hinges depend only upon *the difference between plastic moments and moments computed elastically* (the rotations do not depend directly on the external loads, but indirectly, through the two bending moment diagrams – elastic and plastic). Closer are chosen plastic moments M_p to the elastic moments (under the same external load) smaller are the plastic rotations.

If the chosen plastic moment diagram matches exactly with the elastic one $(M_p = M_{el})$, then $\theta_p = 0$. If the plastic moments differs from those computed elastically, plastic deformations would be developed at a certain level of external loading. Accordingly, more opened cracks will occur in these zones, corresponding to the yielding of the tensioned reinforcement.

The disadvantage of excessive (but limited) local cracking can be accepted when the chosen plastic moments lead to evident economical and/or constructive advantages. Such advantages are related with avoidance of reinforcement agglomeration in certain zones, with a more rational repartition of rebar along the elements, and with capitalization of resistance reserves of redundant structures.

It was experimentally found that the plastic deformations are under control (no excessive cracks in plastic zones, or plastic rotations) if *the plastic redistribution* $\overline{M} = M_p - M_{el}$ is limited to 30% from M_{el}.

3.3 Post-Elastic Analysis of Reinforced Concrete Structures through Adjustment of Elastic Moments

The conclusion of the precedent paragraph, referring to the control of the plastic deformations by limiting the plastic *redistribution* $(M_p - M_{el})$, suggests a practical approach, very simple, for the post-elastic design of reinforced concrete structures.

The basic idea is that, within *the plastic design problem* (see paragraph 3.2), the design moments can be chosen different from the elastic ones. In order to keep the plastic deformations within acceptable limits, the difference between the moments computed elastically and the plastic moments ("plastic redistribution") should be limited to 30%. We accept the disadvantages (even limited) generated by the plastic deformations if the plastic moments diagram brings out real constructive and economic advantages.

The procedure that results from the above considerations is the following:

(i)   Internal forces generated by the external loads (ultimate value) are computed with a standard elastic analysis procedure;

(ii) Plastic moments M_p that allow proportioning the concrete sections and to determine the reinforcement amount and detailing, in n critical sections are chosen, different from the elastic moments. The plastic moments are chosen so that evident advantages from the point of view of uniformity of the reinforcement areas, avoid of reinforcement agglomerations, limit the excessive internal forces at certain critical sections (as for example limiting the shear force in the structural elements by limiting the plastic moment in the sections at the ends of the elements) are obtained. The plastic moments will not differ by more than 30% from the elastic ones;

(iii)   Bending moments in all critical sections of the structure are determined with the aid of simple equilibrium equations (using, generally, the principle of virtual work).

(iv)   In order to ensure that the plastic moment diagrams are compatible with the basic assumptions of the post-elastic analysis, adequate constructive provisions should be implemented, so that:

  • *Potential plastic zones* have enough ductility for development of substantial post-elastic deformations.
  • *Premature brittle failure* by shear, buckling of compressed slender members, loss of anchorage of rebars, etc is avoided,
  • The *elastic zones* of the structure are provided with extra strength capacity in order to avoid uncontrolled propagation of the plastic zones.

These requirements should be specifically fulfilled for each structural system. One of the objectives of the book herein is to analyze these aspects for the most frequently used reinforced concrete structures subjected to seismic actions.

Numerical Examples

Numerical Example Nr. 1: Determine the magnitude of ultimate load of the simply supported beam of Figure 3.E.1, using the Post-Elastic Analysis through "Limit Equilibrium" (item 3.2.3).

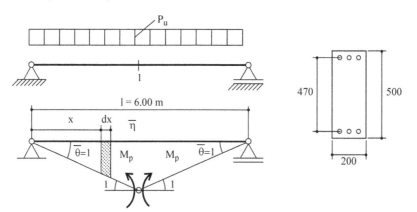

Figure 3.E.1   Failure mechanism and virtual displacements and rotations of a simply supported concrete beam

Reinforced concrete beam has a span $l = 6.0$ m, cross section of $b = 200$ and $h = 500$ mm symmetrically reinforced with $3 + 3$ bars $\Phi = 20$ mm with $f_s = 400$ [MPa].

Its critical section located at the midspan (due to symmetry) has a plastic moment $M_p = 162$ [kNm].

At the ultimate limit state (kinematic mechanism) the load p_u is balanced by the pair of moments M_p developed within plastic hinge. Using the virtual work principle following equation can be written (Fig. 3.E.1).

$L_{ext} = L_{int}$, where L_{ext} is the virtual work of external forces and L_{int} – the virtual work of internal forces (moments in plastic hinge), or (with notations of Fig. 3.E.1):

$$\int\int p_u \bar{\eta} \, dA = \sum M_p \bar{\theta}_i. \tag{3.13}$$

Since

$$\int\int p_u \bar{\eta} \, dA = p_u \frac{l^2}{4}, \quad \text{and} \quad \sum M_p \bar{\theta}_i = M_p \, (1+1), \, results:$$

$$p_u = \frac{8M_p}{l^2} = 36 \text{ [kN/m]}. \tag{3.14}$$

Actually, the example is trivial: structure is statically determined and the moment can be determined through direct moment equation. It was done just for comparing with the below examples.

Numerical Example Nr. 2: Let us consider the same beam as in Numerical example #1 but with fixed ends (Fig. 3.E.2). Suppose that, through appropriate amount of reinforcement but keeping unchanged the concrete section, the plastic moments at critical sections 1, 2 and 3 have same magnitude $M_p = 81$ [kNm] $= 0.50 M_0$ where M_0 is the plastic moment of the simply supported beam of the Example #1. To determine the ultimate load of this beam is required.

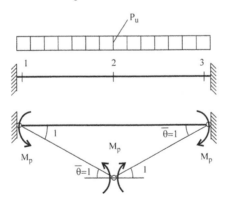

Figure 3.E.2 Ultimate load of fixed ends beam

The failure mechanism comprises three plastic hinges (sections 1, 2, 3), acted by the plastic moments M_p. The work of external uniformly distributed load p_u keeps the same magnitude as in example #1: $p_u l^2/4$, while the work of plastic moments is:

$$\sum M_p \bar{\theta}_i = M_p \, (1+2+1). \tag{3.15}$$

So,

$$p_u = \frac{16M_p}{l^2} = \frac{16 \times 81}{6^2} = 36 \,[\text{kN/m}], \text{same as for Example \#1}. \tag{3.16}$$

Resume the problem considering:

(a)   Plastic moments in sections 1 and 3:
$M_{p1} = M_{p3} = 0.33 \ M_0 = 54 \,[\text{kN m}]$ and in section 2 – $M_{p2} = 0.67 \ M_0 = 108 \,[\text{kN m}]$ and
(b)   Plastic moments in sections 1 and 3:
$M_{p1} = M_{p3} = 0.67 \ M_0 = 108 \,[\text{kN m}]$ and in section 2 – $M_{p2} = 0.33 \ M_0 = 54 \,[\text{kN m}]$, which corresponds to elastic moments.

For both situations we find $p_u = 36 \,[\text{kN/m}]$
Conclusion: For a statically indetermined fixed-ends beam the maximum bending moment corresponding to the simply supported beam can be arbitrarily shared between the critical sections (ends and midspan) provided that the equilibrium with external ultimate load is observed and the critical sections have sufficient ductility to ensure appropriate moments redistribution without premature failure.

Numerical Example Nr. 3: Determine the magnitude of ultimate loading factor λ_u of the structure of Figure 3.E.3, using the Post-Elastic Analysis through "Limit Equilibrium" (paragraph 3.2.3). It is supposed that the plastic moment in critical sections of beams have the same magnitude M_p and at columns' base $M_{p,c} = 2M_p$.

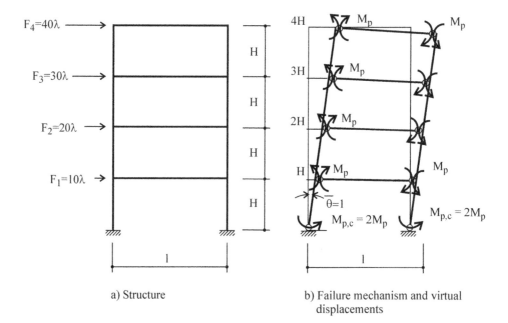

a) Structure   b) Failure mechanism and virtual
                                   displacements

Figure 3.E.3 Multistory elastic-plastic frame with horizontal variable loading

Virtual work of external forces is:

$$\sum \lambda F_i \, \bar{\eta}_i = (10H + 20 \cdot 2H + 30 \cdot 3H + 40 \cdot 4H)\lambda = 300\lambda \tag{3.17}$$

Taking into account that each plastic moment at the beams' plastic hinge as well as the columns' base rotates with same virtual angle $\bar{\theta} = 1$, total virtual work of internal forces (i.e. plastic moments M_p and $M_{p,c} = 2M_p$) will be:

$$\sum M_p \, \bar{\theta} = 12M_p. \tag{3.18}$$

Consequently, the ultimate magnitude of loading factor will be:

$$\lambda_u = \frac{12M_p}{300H} = \frac{M_p}{25H}. \tag{3.19}$$

Total ultimate horizontal load of the structure, which is the structure lateral load capacity, is:

$$F_c = \sum F_i = 100\lambda = \frac{4M_p}{H}. \tag{3.20}$$

For a structure with $H = 3.00\,\mathrm{m}$ and with beams'-ends plastic moments $M_p = 267\,[\mathrm{kNm}]$ the lateral force capacity would be $F_c = 356\,[\mathrm{kN}]$.

Numerical Example Nr. 4: Determine axial force in each column of the example #3. Shear force of beams'-end loads axially the two columns resulting (Fig. 3.E.4).
  Shear force at beam-end: $V = 2M_p/l$
  Axial force in columns according to the diagrams of Figure 3.E.4.

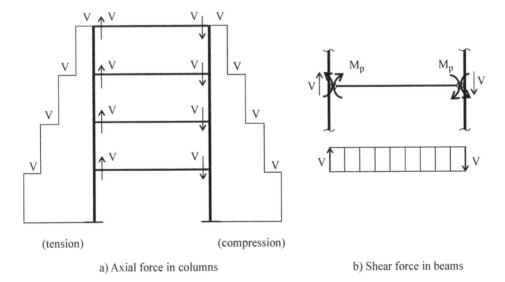

(tension)   (compression)

a) Axial force in columns   b) Shear force in beams

Figure 3.E.4 Shear and axial forces in the frame members

Suggested problem: Using common structural analysis software determine the elastic response of the frame of Figure 3.E.3 under same magnitude of horizontal forces. Assume that $l = 6.00$ m and flexural stiffness of beams are four times higher than that of columns: $(EI)_{beam} = 4(EI)_{column}$. Compare elastic and ultimate state distributions and magnitudes of internal forces.

Conclusions

The specific feature of concrete structures behavior is its non-linear, plastic, character. Accordingly, this chapter of the book examines ways of adapting the general methods of plastic analysis and design to the case of concrete structures subjected to static loadings. This is a first step towards developing concepts and methods for seismic analysis and design of concrete structures, which will be presented within the next chapters.

Chapter 4

Seismic Analysis and Design Methods for Reinforced Concrete Structures

Abstract

This chapter starts with a presentation of the *seismic action* and of *seismic design philosophy* considered to be fundamental premises for development and understanding specific structural analysis and design methods for earthquake prone buildings. Principal consequences resulted from analysis of these premises are further synthesized *in specific requirements for structures subjected to high intensity seismic actions.* Currently used elastic analysis and design approach based on equivalent seismic force is thoroughly examined and commented. Taking explicitly into account the post-elastic character of the seismic response, pushover procedures are a significant step toward its correct modeling. Besides the classical approach described within Chapter 3 advanced methods of seismic pushover analysis are discussed: adaptive pushover procedures, multi-modal analysis, and spatial (3D) pushover analysis. The dynamic post-elastic analysis of single-degree-of-freedom systems and the derived notion *of inelastic spectra* are further presented along with their practical use in seismic analysis and design of reinforced concrete structures. The presentation is further extended to multi-degree-of-freedom systems. Finally, the *modern seismic performance based design* is described and commented upon.

4.1 General Considerations

Earthquakes induce in structures specific effects that substantially differ from those generated by all other actions like gravity or wind loads.

A frequent, grave error in structural design is the extension of design philosophy of gravity-load-dominated structures to those subjected to high intensity seismic actions.

Specific analysis and design methods of structures subjected to high intensity seismic actions are related with two fundamental aspects:

1. Seismic action peculiarities and
2. Specific design philosophy of seismic design which drastically differs from that corresponding to gravity-load-dominated structures

4.2 Seismic Action

Earthquakes are sudden, very irregular, shaking of the earth ground that occurs at certain time intervals, having duration of several seconds to about one or several minutes.

Earthquakes may result from a number of natural or human-induced phenomena (volcanoes, explosions, etc). By far, the most destructive earthquakes are related with relative movement of *tectonic plates* being known as *tectonic earthquakes*.

According to the *theory of plate tectonics*, the rigid earth crust – the *lithosphere* – is divided in number of plates and sub-plates. The plates are floating, very slowly but continuously, on the softer layers immediately below having rheological properties under high temperature and high pressure, called *asthenosphere*. Since the displacement velocity of the plates are not the same (due to inherent non-uniform physical properties – including temperature fields of the asthenosphere), along their boundaries complex contact stress fields occur with predominant compressive, tensile or shearing effects. Obviously, due to plate boundary roughness, the stresses are not uniformly distributed along the plate contour, showing local concentrations. The progressive relative displacement of neighboring plates increases the magnitude of the stresses until the local capacity of the rock is reached. At this moment a sudden relative movement of plates occurs, propagating through surrounding medium as *body waves* or *seismic waves*.

In *energy terms* one can say that, during the relative displacement of plates, the strain energy of the process is stored until a certain threshold, defined by the local energy absorption capacity of the boundaries, is reached. At this moment the stored energy is released and propagated through the body waves to distance of several to some hundreds kilometers around the source.

At the ground surface, the waves are perceived as very irregular, spatial (3D), motions of each point of the soil that constitute the *earthquake* or *seismic event*.

Strong earthquake motions are among the most destructive natural disasters, causing great economic and, sometimes, human losses.

A point called *hypocenter* or *focus* characterizes the rupture surface, which is the source of an earthquake. For small earthquakes the rupture is extended over a limited area and the focus representing the point source is acceptable; for strong earthquakes the rupture involves surfaces of over hundreds of kilometers and the focus (or hypocenter) refers to the zone where the failure was initiated.

The *epicenter* is the point on the earth surface situated on the vertical line that passes through the hypocenter.

Focal distance or *focus depth* is the distance between epicenter and hypocenter.

A considerable progress in understanding the seismic action and its effects on constructions was achieved when *earthquake accelerograms* (time-history), recorded by strong motions accelerographs, became available (Fig. 4.1). Integration of accelerogram gives the velocity and ground displacements as time-history process in an analog or digital format.

Magnitude is a measure of the energy released during an earthquake and, hence, is related with the intrinsic size (power) of the seismic event. The *Richter magnitude* is defined as a function of the maximum ground displacement recorded in standard conditions. The *moment magnitude* is a much more sophisticated notion, based on integration of significant parameters related with the rupture along the fault area.

Intensity is a notion that characterizes the severity of *earthquake caused effects* on a site. The intensity is quantified on a scale on which the estimated effects of the earthquake are plotted, ranging from II (earthquake perceived only by persons at upper floors of buildings) to XII (near total damage) (the Mercalli-Modified or

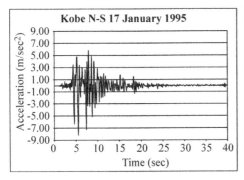
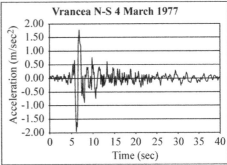

Figure 4.1 Accelerograms of Kobe, 1995 and Vrancea (Romania), 1977 earthquakes (N-S components)

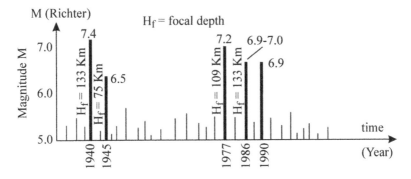

Figure 4.2 Major seismic events in Romania in the last decades (Crainic and Postelnicu, 1992)

MM scales). Since the damage potential of an earthquake depends upon the ground acceleration characteristics (peak ground acceleration *PGA*, duration and frequency content) attempts were made to estimate the intensity through parameters such *PGA* or *effective ground acceleration EPA*.

The intensity of an earthquake on a given site depends significantly upon the distance to the epicenter. This can be quantified by *attenuation relationships* i.e. peak ground acceleration as a function of distance to the epicenter.

Seismic events occur randomly in time (Fig. 4.2). As a general rule, we can say that strong earthquakes are less frequent than earthquakes with lower magnitude. This is expressed through the *return period* of an earthquake of given magnitude (on a given site) which is the average recurrence interval for expected earthquakes with equal or larger magnitude.

Concluding this brief presentation, we can say that the seismic action is globally characterized by following peculiarities:

- It is a *time dependent, transient* (short duration) phenomenon
- The ground shake is *spatial* (three-dimensional)

- The return periods of high-intensity earthquakes depends upon the site. For sites with high seismic activity this can be as much as several decades.

4.3 Seismic Design Philosophy

The structural configuration and proportions, methods of analysis and design are strongly influenced by the building *seismic exposure*. According to this criterion, broadly, two types of buildings exist:

1. Buildings situated in non-seismic or very low seismic areas ("non-seismic buildings") and
2. Buildings subjected to high intensity seismic actions.

There is a substantial difference between *requirements* and *safety factors* corresponding to load combination that comprises or not high intensity seismic action.

The safety factors provided by codes for gravity-loads predominated structures are calibrated so that, under design loads, elastic behavior is ensured, even though limited plastic deformations are accepted. (We consider "limited plastic deformations" those which don't disturb the normal use of building and, consequently, don't require repair). An important consequence of this principle is that the design loads for gravity-loads predominated structures have a higher magnitude that those frequently recorded under current service conditions.

For buildings prone to high intensity earthquakes the codes have to take into account the fact that the intensity of seismic action has a strong variation in time: seismic events with lower intensity are much more frequent than that of severe or destructive earthquakes (see Fig. 4.2). The up-to-date approach of seismic design considers, for a given site with significant seismic activity, more levels of earthquakes in correlation with potential damage of buildings. For example, for different return period following level of earthquakes can be defined:

- *Service earthquake* that occur frequently but with small intensity. No significant building damage occurs during such earthquake.
- *Damage initialization earthquake*; damage occur in non-structural elements (partitions, attics, parapets) as well as small and limited damage (incidentally) in structural components
- *Design earthquake*, with intensity and dynamic characteristics (accelerations, spectral contents, etc.) corresponding to a return period of about 100 to 500 years with a certain (accepted) probability.

Reliable analytical researches demonstrate that, if elastic behavior would be imposed to structures under design earthquake they should resist to horizontal seismic force F_{el} of about $(20–50\%)W_t$ where W_t is the total weight of the mass subjected to seismic movement.

Such an approach is not acceptable because of huge technical and financial effort required. This can be accepted only for some special structures (e.g. nuclear plants or other similar constructions).

So, for economical and technical reasons, design of buildings subjected to high intensity earthquakes should be done with reduced level of seismic forces (in comparison with those required by elastic seismic response).

Generally, the design codes accept:

$$F_{code} = \left(\frac{1}{5} - \frac{1}{3}\right) F_{el}. \tag{4.1}$$

This means that the structures subjected to high intensity earthquakes are designed for seismic forces *dramatically lower* than those corresponding to elastic seismic response. In other words, the buildings have to survive to a design earthquake accepting damage of structural and non-structural elements but in certain controlled zones and having controlled extension. So, according to actual seismic design philosophy, the structures designed to resist earthquakes substantially enter in the post-elastic behavior stage and sometimes are going very close to the collapse during high intensity seismic actions.

There are several reasons that allow accepting this approach:

– Transient character of seismic action: the duration of a seismic event is (in average) of about 30–60 seconds, but the stronger part could be only for several seconds;
– Reversal character of seismic motions so that the seismic inertial forces in one direction are followed by forces acting in opposite sense. Consequently, a tendency of permanent recovering exists during earthquakes.
– The structures can be conceived and designed to be able to dissipate seismic input energy through plastic deformations. According to the modern seismic design, location and magnitude of plastic deformations can be controlled and predicted. Consequently, the damage accompanying plastic behavior can be reliably repaired after each severe earthquake.

However, in order to assure a predictable post-elastic behavior and to limit damage, specific requirements will be observed for structures predominantly subjected to seismic actions.

Obviously, this design philosophy is very different in comparison with that corresponding to gravity predominant buildings for which elastic (or quasi-elastic) behavior is assumed during their whole lifetime.

4.4 Specific Requirements for Structures Subjected to High Intensity Seismic Actions

Since seismic force level accepted by codes are substantially lower than that corresponding to elastic behavior specific requirements should be observed in order to limit damage, to control their position within the structure and to avoid dangerous phenomena leading to potential building collapse. Consequently, beside fundamental requirements of *strength* and *stability*, structures subjected to strong earthquake motions have to fulfill more specific requirements, related with the control of post-elastic behavior during high intensity earthquakes. The damage should be controlled in magnitude and location in order to prevent overall collapse and to allow easy post-earthquake repair.

The specific requirements for structures subjected to strong earthquake motions can be synthesized as follows:

1. Strength and stability;
2. Limitation of lateral displacements ("drift" or "sway" control);
3. Ensure favorable dissipating mechanism;
4. Avoid local brittle failure of sections and elements.

Strength and Stability

This requirement is related with the building *ultimate limit states*. It expresses the condition that, under seismic and permanent loads, no collapse occurs. In other terms, the components of the structure as well as the whole structure do not reach the ultimate state of failure under combined action of gravity and seismic forces.

This condition corresponds to the requirement that the internal forces due to gravity and seismic actions (*maximum demands*) should be less than the minimum capacity of sections and elements within the whole structure:

(Maximum demands) $\leq$ (Minimum resistance)

Fulfillment of this condition can be checked through different approaches:

*   Elastic analysis based on equivalent seismic forces
*   Post-elastic static analysis
*   Post-elastic dynamic analysis.

Although this seems to be the most important requirement, in many cases the size of structural components results from other requirements.

Lateral Displacement Control

The structure responds to the seismic horizontal motion by developing *inertial forces* and *lateral displacements* (or *sway*).

The inter-story relative displacement δ or *drift* causes significant internal stresses to *non-structural* elements especially to the fill-in walls.

In order to limit the damage of non-structural elements, the structure has to possess appropriate *lateral stiffness* so that the inter-story drift δ_i/H_i should be less than an effective admissible magnitude.

$$\frac{\delta_i}{H_i} \leq \left(\frac{\delta}{H}\right)_{adm} \tag{4.2}$$

The maximum amplitude of lateral sway has to be also checked in order to limit the second order moments in columns and to limit the discomfort of occupants during earthquakes.

Favorable Dissipating Mechanism

During high intensity earthquakes plastic deformations are expected to occur even in well-designed structures.

The location of plastic deformations depends upon the relative stiffness of structural components as well as upon their resistance Accordingly, the designer can – through appropriate procedures-impose the location of potential plastic deformations in certain favorable zones. These zones have to be not dangerous for the overall stability of the structure, should be easily found, observed and repaired. Plastic deformations have to be avoided in elements like foundations, columns (vital elements for the overall stability of the structure) and in foundations soil.

Generally speaking, the potential plastic zone location has to be controlled in order to assure a favorable predictable overall behavior of the structure.

This is the requirement of *favorable dissipating mechanism*.

Ensure Local Ductility

Because plastic deformations are expected to be developed within the structure *without collapse*, the potential plastic zones of the structure have to behave in a ductile manner so that local brittle failure is avoided.

Depending upon the type of internal forces acting predominantly on the section reinforced concrete can develop substantial plastic deformation provided that appropriate material and detailing are implemented, good quality of construction is observed, etc. Accordingly, potential plastic zones should be specifically treated taking into account predominant internal forces. Specific measures aimed to fulfill this requirement will be examined for each structural system.

4.5 Analysis and Design Based on Equivalent Seismic Force

The seismic motion of the soil induces to building foundation irregular three-dimensional displacements. The structure responds to this excitation by developing inertial forces, internal forces, deformations and node displacements, all of them being function of time.

For practical purposes, instead of *time-history* response parameters, only *maximum response* is considered to be relevant. According to majority of nowadays codes, seismic response is quantified through *seismic equivalent forces* so that dynamic problem of seismic response is replaced by a static one. In order to derive simplified, easy-to-use, formula for determining equivalent seismic forces a set of supplementary assumptions are accepted that sacrifice the theoretical consistency but lead to satisfactory design rules.

The procedure for determining seismic equivalent forces is based on the modal superposition technique of elastic response in conjunction with simplified elastic response spectra (see Paragraph 4.7) adjusted, in a second step, for taking into account the post-elastic structural behavior. Basically, it involves the following steps:

- Equivalence of elastic multi degrees of freedom (MDOF) dynamic system (corresponding to the multistory building) with single-degree-of-freedom systems, using the modal decoupling (Fig. 4.3).
- Derive simplified relationships for elastic response acceleration spectra
- Adjust the simplified elastic spectra (generally, by dividing with a suitable calibrating factor) for taking into account the post-elastic behavior of the structure and other favorable phenomena.

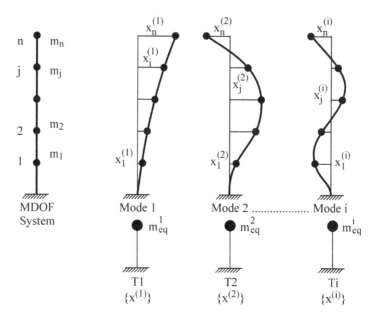

Figure 4.3 Equivalence of MDOF system with SDOF systems

- Calculate the peak lateral force acting on the equivalent *i*-th SDOF system. This is the *total equivalent seismic force* of the MDOF in the *i*-th mode. Calculate the *equivalent seismic force* at each floor (in *i*-th mode) ("vertical repartition")
- Perform elastic analysis of the structure under equivalent seismic forces corresponding to all significant modes, determining internal forces.
- Combine the modal responses using appropriate technique. The most common modal combination method is the *square-root-sum-of-squares* scheme. Note that only the response components (bending moments, axial forces, shear forces, torsional moments) can be combined but not the seismic forces.

The advantage of this conventional method, based on equivalent seismic forces, is its simplicity. With nowadays available software elastic analysis can be easily perform even for complicated 3D structures.

Its main disadvantage is that it is not able to describe *explicitly* the real phenomena that occur during a strong earthquake. All these phenomena are indirectly taken into account through rules. The practical use of the method implies the following steps:

(i)   *Static analysis for gravity loads.* The static model of the structure is initially established, evaluating:

- Geometry (span, height) and topology (elements' inter-connection)
- stiffness of elements (*EI*)

Because the method is conventional, *nominal* sectional stiffness for reinforced concrete members are accepted for example:

- $(0.5–0.6)$ $E_c I_c$ for beams;
- 0.8 $E_c I_c$ for columns subjected to eccentric compression;
- 0.2 $E_c I_c$ for columns subjected to tension

where E_c is the concrete elasticity modulus; I_c – the inertial moment of the gross concrete section.

(ii)   *Dynamic properties of the structure. Dynamic model*, consisting of lumped masses and elastic connections between masses, is defined. The dynamic model can be planar or spatial (3D). The masses involve all structural and non-structural components (all the masses which move due to the earthquake shaking: floors, beams, partitions etc.).

For the dynamic model, the eigenvalues (natural periods of vibration T_i) and the eigenvectors $\{x^i\}$, are determined currently with appropriate software.

For regular, low- or medium-rise buildings only the fundamental mode of vibration is significant and a simplified linear shape of fundamental vector can be accepted.

For each mode, the *equivalence* factor will be determined:

$$\varepsilon_i = \frac{\left(\sum x_j^i m_j\right)^2}{\sum m_j \sum x_j^i m_j} \tag{4.3}$$

The *equivalence factor* in i-th mode allows defining the equivalent mass of SDOF system oscillating in this mode:

$$m_{eq}^i = \varepsilon_i \sum m_j = \varepsilon_i m_{tot} \tag{4.4}$$

(iii)   *Total equivalent seismic force.* Total horizontal seismic force at the building base (i.e. sum of horizontal forces acting in the same direction over the whole building masses, equal to the base shear force of the building considered as a cantilever) (Fig. 4.4) has the general formula (second Newton's dynamic principle):

Force = (Total mass) × (acceleration of the mass) or

$$F_b^i = m_{tot} S_d \quad (i = 1, 2, \ldots, n) \tag{4.5}$$

where: F_b^i is the total (base) equivalent seismic force in mode i,
   m_{tot} – the total mass of the SDOF system which model the building and
   S_d – acceleration design spectrum derived from the elastic spectrum $S_e(T)$ (i.e. maximum seismic acceleration of the elastic SDOF system) divided by a "behavior factor" which expresses the capacity of the system to develop post-elastic deformations.

Different codes accept specific simplified relationships for design spectrum.

(iv)   *Determine the structural elastic response to the equivalent seismic forces.* For this purpose, the total equivalent seismic force is initially distributed to the

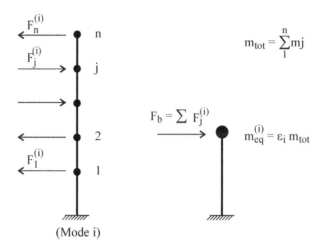

$$m_{tot} = \sum_{1}^{n} mj$$

$$F_b = \sum F_j^{(i)}$$

$$m_{eq}^{(i)} = \varepsilon_i \, m_{tot}$$

(Mode i)

Figure 4.4 Base shear force in *i*-th vibration mode

masses considered by the dynamic model, and then elastic analysis is performed under seismic equivalent forces and permanent loads.

(v) *Modal superposition* of internal forces due to the seismic action (when more than one mode is significant)

(vi) Combinations of the seismic action with other actions according to specific code provisions

(vii) *Design and detailing* of each critical section under internal forces due to gravity and seismic forces.

Comments

Equivalent seismic force approach is the basic *seismic design* procedure for all types of structures. Using appropriate software, space (3D) structures with very few restrictive conditions as well as interaction with infrastructure, foundation soil or infill walls can be analyzed. Associated with *capacity design method* (see Chapter 6) allows proportioning and detailing structural members so that the control of plastic deformations (requirement of favorable dissipating mechanism (see Sub-chapter 4.4)) is fulfilled. However, the user of this procedure has to be aware of its limits and approximations. Actually, even though some corrections can be used in conjunction with this procedure, it remains the most simple and simplified approach. For high performance or complicated structures advanced methods have to be used.

4.6 Post-Elastic Static (Pushover) Analysis to Seismic Actions

4.6.1 General Considerations

Taking explicitly into account, within the analysis, the plastic deformations which accompany the structural response to seismic action pushover procedures is a significant step toward a correct modeling. This can be done at different levels of accuracy according to the relevant aspects required by design process. Besides the classical approach described within Chapter 3 advanced methods of seismic pushover analysis do exist nowadys.

Static (pushover) analysis quantifies the structural response under progressive, monotonic loading which models the seismic action.

Basically, two approaches exist:

- Loading with incremental horizontal *forces* distributed similar to seismic inertial forces and
- Loading with incremental *imposed displacements*.

The method allows determining the *seismic capacity* of the structure, including the progressive development of plastic hinges (location, plastic deformation), potential failure of certain elements and other similar effects and response parameters. Hence, the procedure models a virtual (potential) behavior rather than the response corresponding to a given seismic action.

The post-elastic static analysis has two main applications (Bracci, J. M., Kunnath, S. K. & Reinhorn, A. M. 1997):

1. In conjunction with *post-elastic spectra* (see chapters below) the post-elastic static analysis can be used to check-up the fulfillment of basic design requirement

   Demand ≤ Resistance

   The post-elastic spectra supply the *seismic demands* for a structure in terms of *displacements* or *forces*, while the post-elastic static analysis quantifies the *structural resistance*
2. The post-elastic static analysis is a powerful tool for quantifying the *seismic performances* of a structure. They are required, especially, for the *performance based seismic design* (see Sub-chapter 4.9).

There are more practical procedures for pushover analysis depending upon the constitutive law accepted for critical sections of the structure, the considered loading type (horizontal forces or imposed displacements), the accepted distribution for loading at each step, etc.

4.6.2 "Classical" (conventional) Pushover Seismic Analysis

The procedure presented below is an adaption for seismic actions of the procedure described in paragraph 3.2.2. It accepts the following assumptions:

- Post-elastic behavior of critical section is described through elastic-perfectly plastic relationships. Accordingly, the post-elastic deformations are concentrated in *plastic hinges* (see paragraph 2.4.3).
- Flexural capacities of critical sections are known
- The structure is loaded with gravity forces (permanent and live) considered to be *constant* and with horizontal *incremental* forces distributed over the structure

height similar to the equivalent seismic forces. At each loading step, the horizontal forces magnitude is proportional with a *loading factor* λ:

$$\{F\} = \lambda\{F_0\} \tag{4.6}$$

where $\{F\}$ is the vector of horizontal incremental forces and $\{F_0\}$ – vector of initial horizontal forces.

The procedure involves the following steps:

(i)   Perform the analysis for gravity loads. The vector of bending moments $\{M_G\}$, due to gravity loads, in critical sections is determined.

(ii)   Perform the analysis under initial horizontal forces, resulting $\{M_S\}$

(iii)   Compute, for each critical section i, the ratios γ_i between bending moments due to seismic forces and the available flexural capacity of the section, after superposing with bending moments due to gravity loads:

$$\gamma_i = \frac{M_{S,i}}{M_{cap,i} - M_{G,i}} \tag{4.7}$$

Remarks:

(i)   Each critical section has *two* flexural resistances corresponding to the moment sign – positive or negative. At each loading step, the calculation takes into account the capacity magnitude with same sign as the moment produced by the seismic action. The *resistance flexural reserve* of each critical section is increased or diminished by the gravity load moments depending upon their sign. The gravity load bending moment $M_{G,i}$ increases the resistance reserve of the section if its sign is opposite to that of resistant moment. Hence the reserve flexural resistance of a section is $(M_{cap,i} - M_{G,i})$.

(ii)   Select maximum magnitude of γ_i ratios:

$$Max(\gamma_i) = \gamma_k \tag{4.8}$$

At critical section k first plastic hinge occurs.

(iii)   Determine the magnitude of horizontal forces that produce the first plastic hinge:

$$\{S^{(1)}\} = (1/\gamma_k)\{S_0\} = \lambda^{(1)}\{S_0\} \tag{4.9}$$

(iv)   Consider a hinge in the section k and, accordingly, modify the stiffness matrix of the structure.

(v)   Go to step (ii).

The process will be continued until the structure becomes a mechanism, i.e. the structure stiffness matrix becomes singular. At this moment, the maximum resistance of the structure for lateral loads is reached.

A more accurate procedure would check, at each loading step, the plastic rotations and the shear capacity of each critical section, as compared with admissible

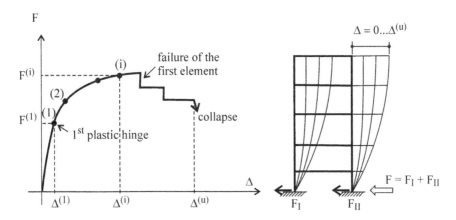

Figure 4.5 Seismic capacity curve for loading with imposed lateral displacements

values. When the admissible values of plastic rotation and/or shear capacity are reached, element failure is expected and the corresponding horizontal load represents the *structure capacity*.

The most common presentation of the pushover analysis results is by plotting the total lateral force at each step (i.e. total seismic shear) versus horizontal displacement at the top of the structure, called *seismic capacity curve*. It can be remarked that, according to the previously described procedure, the post-elastic static analysis is equivalent to a repeated elastic analysis, for the structure having progressive number of hinges.

Comments

a) The procedure above described considers at each step the horizontal loading *increment*. Accordingly, the stiffness matrix of the structure at each step is a *tangential matrix*. The sectional internal force at each loading level is the sum of increments in previous steps.

b) Similar procedure can be used for loading with imposed horizontal displacements having same vertical distribution (over the building height) as that generated by the seismic action. Loading with imposed displacements evidences the *descending branch* of the $F-\Delta$ curve (Fig. 4.5). This quantifies the residual resistance capacity of the structure after gradual failure of certain members. Note that the total lateral force, in this case, is the total horizontal reaction at the base of the structure.

c) For obtaining more reliable information about the seismic behavior of the structure some codes recommend to apply at least two distributions of the lateral loads: the "modal" pattern above described and a "uniform" pattern, with lateral forces proportional to mass at each building level. Parametric studies have shown that for low- and medium-rise buildings the triangular distribution of lateral forces leads to capacity curve which is a good approximation of that obtained through dynamic post-elastic analysis, while for high-rise building the uniform pattern constitutes a better approximation.

An excellent presentation of limits of conventional pushover analysis and of up to date efforts to develop advanced static non-linear analysis procedures can be found in Pinho et al, 2005.

4.6.3 Advanced Procedures for Pushover Seismic Analysis

The "classical" approach of pushover analysis is based on constant (during the analysis) of external loads distribution as well as of magnitude of sectional capacities and members' stiffness. The members' constitutive law accepted is the ideal elastic-plastic model with point plastic hinges. Consequently, the external action and the response parameters depend upon a single parameter: the loading factor.

In order to improve the accuracy of the method considerable research activity has been done. The main efforts have been directed towards:

- Implementing more accurate constitutive laws based on distributed plasticity
- Updating, at each loading step, the response parameters namely: sectional capacities, stiffness, load distribution according to the variation of modal shapes, etc.
- Quantifying the contribution of higher modes ("multi-modal pushover analysis")
- Extending the procedure to space (3D) structures
- Overcoming numerical difficulties which can arise in certain loading domains
- Expressing explicitly the seismic performances of the structure (see paragraph 4.9)
- Identifying and evidencing the sequence of damage state at each loading step (cracking, flexural reinforcements' yielding, members' shear failure, collapse of certain structural or nonstructural components, etc.).

Most relevant results of research activities have been implemented into computer programs like SeismoStruct (SEISMOSOFT, 2008).

A. *Adaptive Pushover Analysis*
Substantial enhancing of pushover analysis has been achieved by updating, at each loading step, the structural response parameters as follows (Papanikolaou, V.K., Elnashai, A.S. & Pareja, J.F. 2005):

a) Due to progressive loading, variation of members' axial force occur and, consequently, the members' flexural capacity changes. This has to be taken into account explicitly since the members capacity determines the sequence of structural plasticization;

b) The progressive loading of structural members generates change of their stiffness (due to extension of plastic deformations) and, accordingly, modal distribution of lateral forces has to be modified;

c) During the progressive loading maximum post-elastic rotation or the shear capacity of certain members can be reached and, so, these members collapse (have to be eliminated from the structure). Accordingly, the structure reaches a new configuration quantified by a new stiffness matrix. This corresponds to a sudden drop in structure capacity. This phenomenon can be evidenced only when the structure is loaded with imposed displacements. Loading of the structure can be further performed if its overall stability is ensured even without eliminated elements.

When the ultimate state (total collapse) is reached the structure stiffness matrix becomes singular (degenerated) and the corresponding total horizontal force is the *structure seismic capacity.*

The analysis performed under above conditions is called *pushover adaptive analysis.*

B. *Multi-modal Pushover Analysis*

For medium- and high-rise buildings more modal shapes and participation factors have to be considered in order to obtain a more accurate seismic response.

Thus, the overall structure response will include more relevant modes.

The first attempt to take into account higher modes within pushover analysis was by performing conventional pushover analysis (force or displacement controlled) for each considered mode and then to compose the modal responses, scaled by appropriate participation factors, through a "modal combination rule". Whenever modal responses may be regarded as independent of each other the maximum value of a seismic action effect (internal forces or displacements) $E_{\max}$ may be taken as: $E_{\max} = [\Sigma E_i^2]^{1/2}$ ("SRSS" rule i.e. square-root-sum-of-squares), where E_i is the value of considered seismic action effect due to the vibration mode i.

More refined approaches are based on the principle of adaptive pushover analysis updating at each loading step the lateral load distribution according to the modal shapes and participation factors. Nowadays computer software has implemented such approaches (Chopra, A. K. & Goel R. K. 2004).

C. *Spatial (3D) Pushover Analysis*

Buildings which don't comply with regularity criteria can no longer analyzed using a planar model. In this respect, we say that buildings with L, T, H, I plan shapes have in-plan non-regularities. Non-regularity criterion applies also to buildings with substantial in-elevation setbacks. Criteria for regularity (or non-regularity) are specifically quantified by different seismic design codes.

Buildings with non-regularities respond to seismic action by developing (besides translations according to principal axis) significant torsional effects.

Post-elastic seismic analysis on spatial models, under most general assumptions, is a tremendous task. In order to make the problem affordable simplifying assumptions have been accepted.

For structures with flexible horizontal diaphragms, translation modes are coupled with local and general torsional ones. A simplified approach accepts that the horizontal diaphragms are infinitely rigid. Consequently, the seismic movement of each floor of the building is a rigid body one with three significant displacements: translations about two in-plan principal axes and rotation about a vertical axis (general torsion or twist) which passes through the *stiffness centre.* In single story buildings the stiffness centre is defined as the centre of the lateral stiffness of all seismic resistant components. In multi-story buildings only approximate definitions of the stiffness centre are possible. The simplified definition assumes that all lateral load resisting systems are continuous over the building height and their deflected shapes under horizontal loads are not very different.

Based on these assumptions easy-to-perform simplified procedures have been developed adapted for "by hand" analysis or for preliminary design.

Up-to-date software packages are able to perform spatial pushover analysis under most general assumptions which takes into account diaphragm flexibility, members' distributed plasticity as well as the step-by-step modification of spatial modal shapes within progressive incremental loading according to general adaptive procedure.

4.7   Dynamic Post-Elastic Analysis of Single-Degree-of-Freedom Systems. Inelastic Spectra

Post-elastic dynamic analysis is considered to be the most advanced tool which models the structural response to the seismic action. It explicitly takes into account the dynamic (time-dependent) character of both – action and response – and the post-elastic behavior of the structure. However, in order to be effectively used in structural design, a good understanding of its basic assumptions, methodology and limits is required.

Let us consider a simple structure – a cantilever with a single mass m at top, having a planar horizontal movement and rotational energy of the mass that can be neglected (Fig. 4.6). This structure can be modeled as a *single-degree-of-freedom system*.

The cantilever behavior under horizontal loads is modeled through appropriate constitutive law (Fig. 4.7).

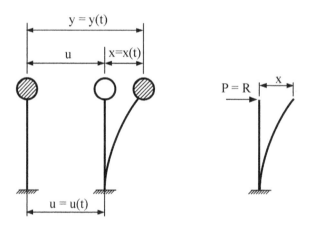

Figure 4.6   Single-degree-of-freedom system

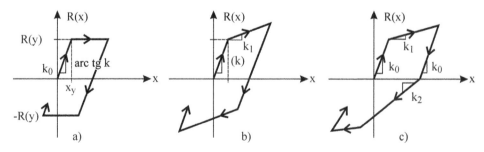

Figure 4.7   Simplified constitutive laws for a cantilever with single lateral force a) ideal elasto-plastic, b) elastic-plastic with strain hardening, c) elastic-plastic with stiffness degradation

Each constitutive law is analytically expressed through appropriate functions. For example, the elastic-ideal-plastic law (Fig. 4.7a) is defined by the function:

$$R(x) = \begin{cases} k_0 x & \text{if } x \leq x_y \text{ and } \dot{x} > 0 \\ R_y & \text{if } x > x_y \text{ and } \dot{x} > 0 \\ -k_0 x & \text{if } x > x_y \text{ and } \dot{x} < 0 \end{cases} \qquad (4.10)$$

During earthquakes, the structure base follows the ground motion $u(t)$. Each one-direction component of the ground motion is described through digitized records expressed, generally, in terms of accelerations (i.e. *accelerogram*): $\ddot{u}(t)$.

Cinematically, the mass motion can be described through one of the following parameters:

- Total (absolute) displacement $y = y(t)$
- Relative displacement $x = x(t)$.

We have:

$$y = x + u \qquad (4.11)$$

The seismic dynamic response of the system can be determined using the d'Alemebert principle. According to this principle, the system of forces effectively acting on the mass m, during its movement, including the *inertial force* $F_i = -m\ddot{y}$, the *viscous damping force* $F_a = c\dot{x}$ and the *restoring force* $R(x)$ is equivalent to zero ("pseudo-dynamic equilibrium"), hence:

$$m\ddot{y} + c\dot{x} + R(x) = 0 \qquad (4.12)$$

Since $\ddot{y} = \ddot{x} + \ddot{u}$, the equation can be written as

$$m\ddot{x} + c\dot{x} + R(x) = -m\ddot{u} \qquad (4.13)$$

which is the *equation of motion* of a single mass supported by a cantilever having a known constitutive law $R = R(x)$, subjected to a base displacement defined through a given (digitized) accelerogram. This is a differential non-homogenous, non-linear equation of second degree with constant coefficients. The non-linearity results from the term $R(x)$.

The equation of motion can be solved through numerical, step-by-step in respect to time, procedures (like Newmark's or Newton-Raphson procedure). Its solution is the *seismic response* of the system quantified, in a *time-history* format, through kinematic parameters:

- Relative mass displacement $x = x(t)$
- Relative mass velocity $\dot{x} = \dot{x}(t)$
- Absolute acceleration $\ddot{x} + \ddot{u} = \ddot{x}(t) + \ddot{u}(t)$.

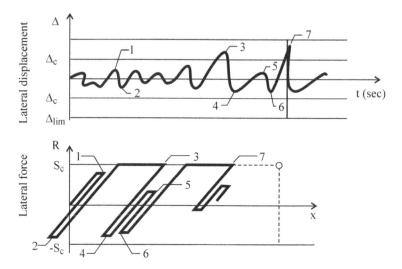

Figure 4.8 Hysteretic loops of a single-degree-of-freedom system

Plotting against the time the response of a system with given strength capacity R_y to an earthquake motion with a known accelerogram $\ddot{u}(t)$ a picture of the most relevant phenomena induced by the seismic action to the system is obtained.

The mass displacement x which exceeds the maximum elastic one x_y quantifies the *duration* and *time of occurrence* of the *post-elastic excursions* of the system.

The ratio between maximum displacement and that corresponding to yielding is the *ductility demand* of the system for the considered earthquake:

$$\mu_{\text{nec}} = \frac{x_{\max}}{x_y}.\tag{4.14}$$

If the restoring force R is plotted against the mass displacement x, the hysteretic loops are evidenced (Fig. 4.8). The total area of the hysteretic loops gives the magnitude of the *energy dissipated through plastic deformations* of the structure.

Inelastic Spectra

For practical purposes, generally, only the maximum response is required.

The maximum post-elastic response *to a given seismic input*, expressed in terms of displacement, velocity or acceleration, significantly depends upon the following factors: natural period of the elastic oscillations T, nominal system damping, the *shape* of the constitutive law, and the limit of the elastic domain expressed through the maximum elastic displacement x_y or through the corresponding lateral force R_y.

If the maximum absolute post-elastic response is plotted versus the natural period T, *inelastic response spectra* are obtained (Fig. 4.9). In Figure 4.9 the parameter used

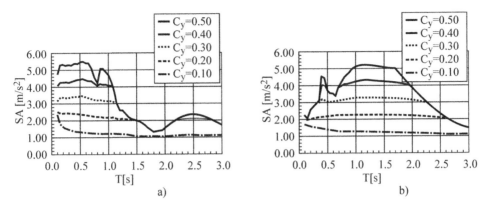

Figure 4.9 Inelastic response spectra for a) El Centro 1940 and b) 1977 Earthquake, recorded in Bucharest (North-South components) (courtesy of I. Craifaleanu)

for different spectral relationship is $C_y = F_y/W$, where F_y is the lateral yielding force and W the mass weight.

With appropriate computer software, the inelastic response spectra can be easily obtained for any seismic input expected on a site, and for constitutive laws that characterize the behavior of different structural materials and members. They enable to predict:

- Maximum lateral displacement induced by the seismic motion $|x|_{max}$, called also *target displacement*
- Maximum inertial force developed by the elastic-plastic system during earthquake.

However, some comments are necessary for understanding the use and the limits of inelastic spectra approach.

1. The *inelastic design spectra* have to consider different potential seismic inputs (accelerogram) which comply with local site conditions, and different (possible) constitutive laws. Accordingly, envelopes of several potential seismic inelastic spectra have to be taken into account.
2. The maximum inertial force developed in an elastic-plastic system during earthquake is, actually, limited by the *strength capacity* of the structure (maximum structure resistance) $F_{max} \leq R_y$.
3. On other hand, the maximum inertial force for a single-degree-of-freedom inelastic system can be calculated as

$$F_{max} = m|\ddot{x} + \ddot{u}|_{max} = mS_a^{in} \tag{4.15}$$

where S_a^{in} is *the inelastic acceleration response spectral magnitude*.

4. By comparing F_{max} with the *seismic equivalent force* relationship (see above paragraph), for single-degree-of-freedom system, one can see that the seismic

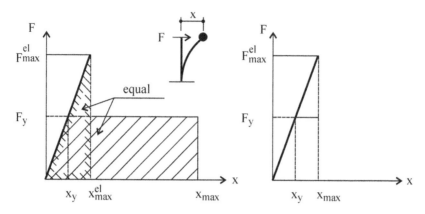

Figure 4.10   Maximum seismic displacements for a) short- and b) long-period structures

equivalent force could be more exactly calculated from *averaged* smoothed *inelastic* acceleration spectra.

5.   By comparing elastic and post-elastic seismic response for different input accelerogram as well as for different constitutive laws one can see that:

- For *short period* structures the *inelastic spectral displacement* is close to that of an *elastic* system that has same *energy absorption* as the inelastic one (i.e. same area under force-displacement curve – see Fig. 4.10a). It follows that, for these systems, the post-elastic inertial force can be deduced from maximum elastic inertial force divided by

$$q = \frac{1}{\sqrt{2\mu - 1}} \qquad (4.16)$$

(μ is the member ductility factor $\mu = \frac{x_{max}}{x_y}$)   (4.17)

This statement is known as *equal-energy principle*.

- For *long period* structures the *inelastic spectral displacement* is close to the maximum displacement of an *elastic* system that has same natural period (Fig. 4.10b):

$$S_d^{in} = |x^{in}|_{max} = |x^{el}|_{max} \cong S_d^{el} \qquad (4.18)$$

Accordingly, the post-elastic maximum seismic force results from maximum *elastic* force divided by

$$q = \frac{1}{\mu}. \qquad (4.19)$$

This rule is known as equal-displacement principle.

These statements cannot be rationally demonstrated; they result from parametric studies involving several seismic inputs and different constitutive laws for the structure.

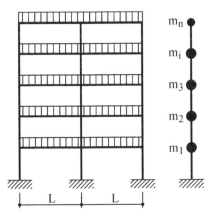

Figure 4.11   Planar frame modeled as multi-degree-of-freedom system

4.8   Dynamic Post-Elastic Analysis of Multi-Degree-of-Freedom Systems

The above discussion can be expanded for *multistory structures* that can be represented by *multi-degree-of-freedom (MDOF)* dynamic systems (Fig. 4.11) (Chopra, 2005).

For multi-degree-of-freedom systems the seismic response results by solving a set of simultaneous differentiate equations which quantify the system equilibrium wrote, normally, in a matrix format. Numerical step-by-step, in respect with time, procedures as well as corresponding computer software are nowadays available for integration of these equations (Bathe, 1982). Time-history programs like those included in *ETABS, DRAIN, IDARC* software packages are among the most used for this purpose.

For each such program the *input data* are:

- Geometry and topology of the structure
- Elastic stiffness of each structural component
- Lumped mass at each node
- Shape of constitutive law for structural components and its characteristic parameters (i.e. yielding forces, maximum elastic displacement, etc.)
- Seismic input given as *digitized accelerogram.*

The *time-history output* consists of following quantities:

- Internal forces at critical sections (i.e. bending moment, axial force, shear force, torque)
- Node lateral displacements
- Inter-story drifts
- Absorbed and dissipated input energy

These magnitudes are the closest approximation of the real time-history seismic response of post-elastic structures under a given accelerogram. Accordingly, the post-elastic analysis can be regarded as the most accurate tool for checking the fulfillment of structural seismic requirements.

4.9 Performance-Based Design

As discussed in previous paragraphs, according to the seismic design philosophy of current codes, the main objective of up-to-date design of buildings subjected to severe earthquakes is to ensure *life-safety* i.e. to ensure that the buildings *survive* to design earthquake. Damage of structural elements (or, in other terms, post-elastic behavior) is accepted, but within certain controlled zones and with limited extensions.

The second objective of seismic design codes is to *avoid structural damage* and to *limit damage of non-structural elements* for small and moderate earthquakes. This requirement is taken into account through some constructive rules and through calibration of safety coefficients rather than explicitly considered within the design process.

This design philosophy is based on observations of effects of past earthquakes and on extensive researches, both theoretical and experimental, performed during several decades.

The effectiveness of actual codes has been verified during severe earthquakes occurred in past decades.

It was proved that the *life-safe* goal was reliably achieved for buildings correctly designed according to modern seismic codes.

On the other hand, the economic losses due to damage of non-structural elements and to interruption of building function appeared to have an unacceptable extent. For instance, for the M = 7.1 Loma Prieta Earthquake (San Francisco Bay area; October 17, 1989) the estimated losses were more than \$7 billion, while for Northridge Earthquake (M = 6.9) of the January 17, 1994 the losses were evaluated at approximately \$20 billion.

It became obvious that a new generation of seismic codes was needed, in order to implement a *comprehensive approach* aimed to minimize the expenses related with occurrence of severe seismic events.

The Structural Engineers Association of California (SEAOC) undertook this task, establishing in 1992 the *Vision 2000 Committee*. The Vision 2000 Committee defined a framework for further development of a new generation of seismic codes. This is contained in the two-volume document *Performance Based Seismic Engineering of Buildings* (SEAOC Vision 2000 Committee, 1995). It was expected that the new set of seismic design codes (involving regulations for design of new buildings as well as for evaluation of seismic resistance and for re-design of existing buildings) would lead to obtaining buildings with predictable response and damage extent for different seismic events levels. The completion of this task involved a massive and lengthy effort. The Federal Emergency Management Agency (FEMA) funded this project.

The basic concept of the new approach is that of *seismic performance*.

In general terms, the seismic performance is an index of the overall building behavior when subjected to an earthquake of certain intensity.

In technical terms, the *seismic performance* is related to the *damage* suffered by a building under a certain seismic level, involving structural and non-structural components, as well as the contents. A *seismic performance level* is a *limiting damage state*.

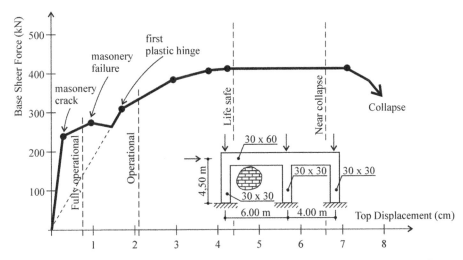

Figure 4.12 Force-displacement diagram (capacity curve) illustrating the seismic performance levels of a simple structure

Quantitatively, the seismic performance levels of a building can be quantified through the magnitude of the building's *lateral displacement* generated by the seismic action.

These definitions and statements can be illustrated on a force-displacement diagram of a simple reinforced concrete frame (structural component) with a masonry infill panel (Fig. 4.12).

The seismic in-put level can be quantified through the magnitude of the lateral force or of lateral displacement. According to the shape of the force-displacement curve, behavior stages corresponding to different lateral force magnitude (i.e. seismic input levels) can be identified: elastic response of both frame and infill, cracking initialization of the masonry infill, infill collapse, further elastic behavior of pure frame, occurrence of first plastic hinge (i.e. structural damage initialization), post-elastic frame response and, finally, frame failure.

Although the continuous shape of seismic force-lateral displacement curve suggests an infinite number of performance levels corresponding to the infinite number of sequences of the seismic event, for practical purposes a discrete number of building performance levels are selected.

For each earthquake design level, a damage extent is accepted taking into account factors like building occupancy, function, total costs of building construction, maintenance and post-earthquake repair, etc.

According to the genuine version of this approach, defined by the SEAOC Vision 2000 document, standard *performance levels* are as follows:

A. *Fully operational.* An elastic response is expected with no or minor damage. Post-earthquake, the building and its equipment and services can be immediately used at full capacity and in safe conditions.

Table 4.1 Earthquake Design Levels

Earthquake Design Level	Recurrence Interval	Probability of Exceeding
Frequent	43 years	50% in 30 years
Occasional	72 years	50% in 50 years
Rare	475 years	10% in 50 years
Very Rare	970 years	10% in 100 years

B. *Operational.* Moderate damage to non-structural elements and contents and light damage to structural elements is recorded post earthquake. The building is available for occupancy although some repair or use of back-up systems could be necessary for its full function.
C. *Life-safe.* This performance level is characterized by moderate damage to structural and nonstructural elements. The structure lateral stiffness is reduced; ability to resist supplementary loads is limited but some margin against collapse remains.
D. *Near collapse.* This is an extreme damage state in which the resistance of the structure has been substantially compromised so that aftershocks can induce building collapse.

The performance levels are quantitatively related to the *lateral displacements* (or drifts) rather than to *lateral forces*. (Generally, in post-elastic range, the forces are not suitable for describing the structure response, since the force-displacement relationship is flat so that small variation of force corresponds to large displacement variation). Accordingly, the seismic performance based design is sometimes defined as *displacement based seismic design.*

The design methodology based on seismic performances (or on lateral displacement) presumes that the building performance can be reliably predicted for a given earthquake ground motion. Although advanced analysis methods and corresponding computer software are nowadays available for this purpose, the accuracy of results is still questionable and subject to number of uncertainties. Accordingly, the whole approach as well as limits values for drift, plastic rotation and other characteristic parameters have to be used with caution and permanently subjected to judgment.

For a given location, the earthquake severity is a function of its *return period.* Accordingly, the *earthquake design levels* are expressed in terms of a *mean recurrence interval* or of the probability of exceeding (Table 4.1).

The desired seismic performances should be related with the expected earthquake design level. The design performance expected level to be reached by a building subjected to a seismic event characterized by a certain design earthquake level is a *design performance objective.* The design performance objectives are selected according to the building's occupancy, the importance of the functions sheltered by the building, economic considerations (including costs of post-earthquake repair or business interruption) and potential importance of the building as historical or monumental site.

Minimum performance objectives for *basic facilities* (buildings which are not under special requirements) recommended by Vision 2000 document are listed in Table 4.2

Table 4.2 Recommended performance objectives for basic facilities

Earthquake Design Level	Minimum Performance Level
Frequent	Fully Operational
Occasional	Operational
Rare	Life-safe
Very Rare	Near Collapse

For obtaining a building that complies with standard performance objectives required by the code or imposed by the designer, post-elastic analysis should be performed for each earthquake design level. Thereafter the magnitude of basic response parameters (internal forces, displacements, plastic rotations, etc.) shall be checked, by comparing with standard permissible limits.

The procedure, above described, was followed by a considerable research and analytical effort about advanced post-elastic behavior and analysis of structures to seismic actions. It triggered unanimous interest of code writers and was implemented in up-to-date most relevant seismic design codes. For example, the European EUROCODE 8 accepts two basic objectives (instead of four): (a) No-collapse requirement (Ultimate Limit State ULS): "structure shall be designed and constructed to withstand the design seismic action without local or global collapse, thus retaining its structural integrity and a residual load bearing capacity after the seismic events" and (b) Damage limitation requirement (Service Limit State SLS): "structure shall withstand seismic actions having larger probability of occurrence than the design seismic action, without occurrence of damage and associated limitations of use". Seismic Performance based approach has been also implemented in documents like NEHRP Recommended Seismic Provisions, being used for calibrating current code procedures. It is expected that, accordingly, the conventional code procedures will lead to structures having more accurate and predictable seismic performances under earthquakes with various severities.

Conclusions

Earthquakes induce in structures specific effects that substantially differ from those considered in design of gravity-load-dominated ones. It is a frequent, grave error the extension of gravity-load-dominated design philosophy applied the earthquake prone buildings. Within the present chapter requirements for structures subjected to high intensity seismic actions are listed, explained and commented upon in relation to specific features of the seismic action. Current and advanced up-to-date methods for seismic analysis and design of concrete buildings are then examined and commented upon.

Chapter 5

Structural Systems for Multistory Buildings

Abstract

In contrast to the case of gravity-dominated buildings, the seismic design process has to consider all structural components which transfer horizontal loads to the foundation soil. They can be grouped in three main, relatively homogeneous, components constituting the building structural system: (a) superstructure (currently called structure), (b) infrastructure and/or foundations, and (c) foundation soil. Within the present chapter it is pointed out that the consideration of the mutual interaction of these three components as well as an organized whole is of crucial importance in correct determining and understanding of the seismic building response. Most commonly used reinforced concrete structures for multistory buildings are compared and their optimum usage domain is specified.

5.1 Definitions

Multistory buildings, especially those subjected to high intensity seismic actions, are among the most challenging tasks for a structural engineer.

Design of a high rise building requires a permanent co-operation between its main designers: architect, structural engineer and responsible for building equipment and facilities. It is a matter of talent, constructive imagination and high level knowledge about the main peculiarities of these constructions.

For an efficient use of specific properties of different materials, in engineered buildings structural and non-structural components are differently treated. Non-structural elements: partitions, finishing, casings, roof, etc. ensure the building function and esthetics. Structural components are those which resist and transfer *actions* (forces and imposed displacements/deformations), to the foundation soil, in order to ensure building's *resistance* and *stability*.

The structural components have different shapes and functions and are made from materials of different nature, having different mechanical properties: reinforced concrete, masonry, ground (for foundation soil), etc. Taking into account the role within their basic function the structural elements are grouped in *organized assemblies*. Even though these assemblies or components have different nature they fulfill the condition of an *organized system* having a unique main function: to carry and transfer actions. It is what we generically call, *structural system*.

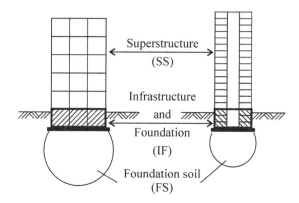

Figure 5.1 Components of structural system

Structural system of multistory buildings consists, generally, of three components: (a) *superstructure* (currently called *structure*), (b) *infrastructure* and/or *foundations*, and (c) *foundation soil*.

Superstructure is the building main resistance skeleton, located at the upper part of the construction, having a relatively homogeneous layout and mechanical properties.

Infrastructure is the bearing system of the superstructure having considerable greater resistance and lateral stiffness than that.

Foundation soil is that portion of the ground which is effective in transferring and spreading the infrastructure mechanical reactions to the surrounding ground, ensuring the overall building equilibrium and stability.

Within structural design process, the consideration of the *mutual interaction* of these three components as well as an *organized whole* is of crucial importance.

5.2 Types of Superstructures

Most commonly used reinforced concrete structures for multistory buildings are of three types:

- *Frames*
- *Structural wall systems* and
- *Dual systems.*

The selection of structural system for a building is primarily determined by the function and architectural considerations, by the total height of the building and by the seismic exposure.

5.2.1 *Reinforced Concrete Frames*

Frames for buildings are spatial (3D) systems of beams and columns rigidly interconnected in joints.

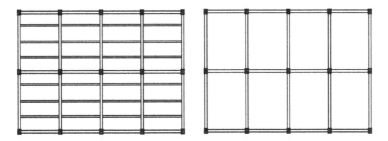

Figure 5.2 General configuration of a frame structure

The structural configuration and proportions, methods of analysis and design are strongly influenced by the building *seismic exposure*. According to this criterion, broadly, two types of frames exist:

- Frames of buildings situated in non-seismic or very low seismic areas ("non-seismic frames") and
- Frames predominantly designed to resist seismic actions.

The design principles and methods of frames subjected to high intensity seismic actions cover a significant part of design problems of high-rise multistory frames located in non-seismic areas.

General Layout of Frame Systems

The most used frame structures have the columns disposed according to the nodes of a rectangular regular mesh (Fig. 5.2).

Span and bay are chosen according to architectural and functional requirements. This means that the cells determined by four columns should fit generally with the requirement of column free room.

Generally, span and bay length are less than 5.00 m for residential buildings. For office buildings or other similar facilities like hospitals, the frames' span is larger and can be as much as 9.00 m or even longer.

There are also special framed structures for show buildings (cinema, theaters, opera) or similar.

When subjected to lateral forces, the building acts as a *vertical cantilever*. The cantilever effect generates important global *shear forces* and *overturning moments*. They act the frame system considered as a whole, which has to transfer them to the foundation soil. Since the most advantageous cross section shape for a flexural element is the *close tubular section,* attempts to obtain a "tube effect" for framed structures have been made. This has been realized providing close spaced columns (at 1.50–2.50 m) along the building perimeter interconnected through very stiff spandrel beams. The resulting structures are *tubular frames* (Fig. 5.3a).

The concept of tubular frame has been extended for super-tall buildings providing a supplementary internal framed (or wall) tube obtaining *tube-in-tube frame structures* (Fig. 5.3b). *Multiple tubes* have also been used (Fig. 5.3c).

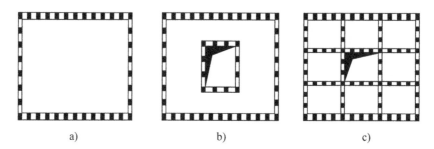

a)   b)   c)

Figure 5.3 Different types of tubular frames

5.2.2 Wall Systems

Multi-story *frame* structures subjected to seismic actions have a limited use for high-rise buildings due to large columns' sizes required at their bottom zone. Consequently, an important part of the building area at bottom floors is not functional.

Search for alternative structural solutions, based on use of vertical elements with high stiffness and resistance, was a condition for extending the reinforced concrete structures for high- and super-high-rise buildings.

One of the most important solution, which in early 1950s, was the use of *structural walls* as an alternative solution to the traditional frame structures. The idea was to convert partitions into *structural walls*. Due to their cross section shape and proportions, the walls have a high "natural" stiffness and a substantial resistance capacity.

Types of Structural Walls

An ideal solution for structural walls is that of compact solid units called *cantilever walls* (Fig. 5.4). Cantilever walls are easy to be executed and lead to structures with unambiguous load paths. The most suitable cross section for cantilever walls is rectangular having or not reinforced boundaries. More solid walls provided on perpendicular directions are sometimes crossed forming elements with different section shapes: T, L or tubular.

In many situations, openings in structural walls are required to accommodate doors or windows. We call the resulting component *structural wall with openings*. When openings are provided in uniform regular pattern over the wall height the unit can be considered to be formed by two or more solid cantilever walls coupled through rigid beams (Fig. 5.5). These are *coupled structural walls*. Besides functional advantages (in comparison with cantilever walls) the coupled walls, appropriately designed and detailed, showed a stable, controllable energy dissipating capacity during high intensity earthquakes. Sometimes, a staggered pattern of opening is preferred. When openings are suitable arranged, with large space between them, the wall has a good stiffness and appropriate reinforcement can be provided to be effective in preventing shear cracks.

Both cantilever and coupled walls can be performed by *cast-in-place* or by *prefabricated* concrete. Inter-connected rectangular prefabricated units called *large panels*, realize prefabricated structural walls.

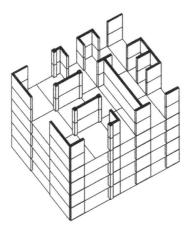

Figure 5.4 Types of cantilever structural walls

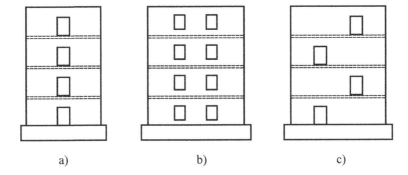

a)   b)   c)

Figure 5.5 Structural walls with openings: a), b), Coupled walls c), Walls with staggered openings

Structural walls supported by columns at the first floor have been used in the past, in order to obtain architectural freedom (large rooms) at this level. These are *walls with weak and soft story.* They evidenced very poor seismic behavior and nowadays are abandoned.

5.2.3 *Dual Systems*

Structures made by frames associated with structural walls are called *dual systems.*
  The dual systems combine the advantages of both frame and structural wall systems.
  Two types of dual systems are defined by codes:

- *Frame-equivalent dual systems* are those for which the shear resistance of the frame system at the building base is greater than 50% of the total shear resistance of the whole structural system and
- *Wall-equivalent dual systems* in which the shear resistance of the walls at the building base is higher than 50% of the total seismic resistance of the whole structural system.

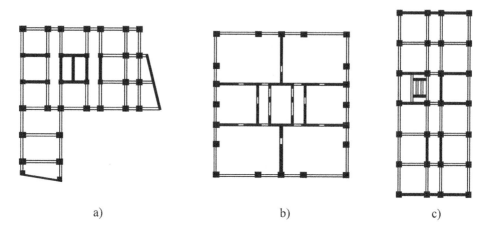

a)   b)   c)

Figure 5.6 Examples of dual systems

5.2.4 *Advantages and Disadvantages. Optimum Usage*

The most significant advantage of frame structures in comparison with structural walls systems is the freedom, from architectural and functional points of view, in space use. The frame structures allow obtaining large rooms free of structural elements. One says that the "flexibility" of the building's *architectural configuration* is the main advantage of the frame structures. This feature is important and useful for hospitals, office buildings, schools, garages but it is less significant for hotels or residential buildings.

Main disadvantages of framed structures are related to the large size of their components – columns and beams – necessary to comply with strength and stiffness requirements specific to tall multistory buildings. This disadvantage limits the use of pure framed structures for high-rise buildings.

Having large cross section area and substantial lateral stiffness the structural walls seems to be an excellent solution for multi-story buildings. However, obvious structural requirements oblige to keep the walls continuous over the building height. Thus, their position is fixed throughout all building stories. Accordingly, the architectural layout is "frozen" over the building height during its lifetime (minor changes can be made only within cells delimited by structural walls). This is a significant restriction for use of structural walls for high-rise buildings. For this reason, the structural wall systems are suitable for apartment buildings or hotels but not for hospitals, offices or other type of buildings which require large *open spaces*.

Combining the advantages of both framed and wall structures, the dual systems considerably extend the use of reinforced concrete for multistory buildings. The key of developing optimum solutions with dual systems is the harmonious working together of architect and structural engineer. Seeking for best building configuration which complies with functional, esthetics and structural requirements, finding appropriate location for structural walls within the building layout, solving correctly the loads transfer to the soil through well-conceived infrastructure are tasks of crucial importance for designers' team of high- and super-high rise buildings. Dual systems are in many cases the best solutions for tall buildings with various destinations.

5.3 Structures with Controlled Seismic Response

According to seismic design philosophy of current practice and codes, based on balance between strength/stiffness/ductility requirements, controlled damage of buildings' structural and non-structural elements under design earthquakes are accepted ("*life-safety*" goal). Even though this goal has been reliably achieved by the buildings designed according to modern codes the cost of losses related with damage repair and with interruption of building function remains considerable. Consequently, development of global strategies for seismic risk management based on new, innovative concepts and technologies became an evident priority.

The main idea, borrowed from industrial and mechanical applications, was to decouple the building from the soil or to increase significantly its damping capacity using mechanical devices added to its structure.

A building subjected to earthquake motions can be considered as a *dynamic system*. Its response to external excitation depends on three major parameters: mass, damping and stiffness. Through appropriate strategy, intervention on any of these parameters can advantageously modify the response. Intervention on parameters which govern the seismic response in order to minimize its effects leads to the concept of *structures with controlled seismic response*.

Solutions for structural control can be grouped in three main categories:

1. Systems of *base isolation*
2. *Damper* systems and
3. *Systems with tuned mass*.

5.3.1 *Systems of Base Isolation*

Seismic base isolation is one of the most frequently used technique of structural control. The idea is to "split" the structural system vertically into parts decoupled from each other through special devices, so that the seismic movement of the foundation is no longer transferred to the upper part (or parts) of the building. The procedure uses different types of seismic isolators.

The isolators are bearing devices which have low lateral stiffness. They allow substantial horizontal displacements (about 20 to 30 cm) and can develop restoring forces. Accordingly, the upper part of the building will respond to the horizontal seismic excitation through quasi-rigid body displacements, with only minor distortions. So, substantial seismic damage will be prevented.

Numerous structural control devices have been imagined, then laboratory tested and, finally, implemented into new or existing structures.

Some frequently used seismic isolators will be briefly described below.

*   *Elastomeric Isolator (laminated rubber bearing)* is a low cost isolation solution, having simple constitutive laws and, accordingly, simple analysis model, easy to understand and implement. Their main disadvantage is related with low damping ratio (about 2–3%) which involves large displacements, causing possible second-order (stability) effects.
*   *Seismic Isolators with Lead Rubber Bearings (LRB)* are similar to elastomeric bearings but provided with a lead core which dissipates energy due to plastic

deformations. Their damping ratio is high, up to 30% of critical damping, and show a good restoring capacity due to their elastomeric component.

- *Sliding bearing devices with controlled friction* – are provided with two metal components, in contact with each other, which dissipates energy through dry friction (Coulombian energy dissipation). An example of such device is that called *friction pendulum system* (FPS). They have the advantage of a pure mechanical system with mechanical properties which don't depend upon the time. The damping capacity is high (10 to 50% of critical damping ratio).

5.3.2   Damper Systems

These solutions imply devices which significantly increase the building damping capacity. Dampers reduce vibration amplitude similar to the hydraulic shock absorbers of an automobile. Damping devices could be *passive*, with constant mechanical properties, or with variable characteristics (*tuned dampers)* adjustable in function of the seismic excitation.

Passive systems are hydro-mechanic devices similar to those used for automobiles, guns, etc. They are triggered by the seismic input signal and their characteristics remain constant during building lifetime.

Active systems adapt their action on the building according to seismic excitation in order to optimize the effect on the building response. They generate additional forces acting onto the structure being strictly related to an exterior energy source.

As an alternative, semi-active systems combine the advantages of passive and active devices, by providing adaptive damping according to seismic in-put magnitude. They improve the seismic response with minimal exterior energy supply. Even in absence of external power these devices perform a passive control.

Tuned damper systems – both active and semi-active – use full automatic devices, computer controlled, according to the seismic in-put characteristics supplied by specific sensors.

5.3.3   Systems with Tuned Mass

These systems act on a building through an additional mobile mass (generally at roof level) generating forces of opposite sense to those induced by the seismic action. The existing applications refer mainly to high rise buildings subjected to wind actions (example: John Hancock Building in Boston, Ma.) but applications for seismic prone buildings do exist too.

The systems with controlled response are an extremely promising solution from the building behavior point of view. They open the perspective of obtaining *intelligent constructions* able to adapt their response to external actions so that the potential damage is minimal or even avoided.

5.4   Infrastructure

When subjected to lateral forces the whole building acts as a *vertical cantilever*. Thus, the structure, considered as a whole, respond to the seismic action by important *overturning moments* and *shear forces*. In order to transfer to the soil earthquake-generated reactions as well as the gravity loads the whole structure has to be provided, at its base, with a bearing system.

The structural system component that transfers to the soil the seismic overturning moments and shear forces, resulted from the "cantilever effect", as well as the gravity loads of the whole building is called *infrastructure*. The infrastructure can be provided (not always necessarily) with local *foundations* that ensure direct contact with the soil.

Generally, infrastructure is identified as the structural component which has a considerably greater *lateral stiffness* and *resistance* than the superstructure.

The infrastructure is currently provided at basement level. Sometimes the infrastructure is extended over the first floor(s) of the building too.

The infrastructure stiffness and resistance is generally obtained through addition of supplementary structural walls at the basement level. Other solutions can also be implemented like a system of *braces* on the first floor of the building.

Due to the change in stiffness at the infrastructure level (in comparison with stiffness distribution above this level) substantial redistribution of internal forces among different components occurs. The total lateral force acting at interface superstructure/ infrastructure is shared to the infrastructure components proportionally to their stiffness. The horizontal diaphragm just above the infrastructure has to ensure the transfer of total lateral force to the infrastructure components.

Specific recommendable solutions for infrastructure of different types of structures (framed, wall or dual systems) and their analysis, design and detailing features will be presented within specific chapters of the book.

As a general rule, robust detailing solutions have to be chosen for the infrastructures since its analysis involves many uncertainties.

Conclusions

Within the present chapter the need of considering the structural system, with its three components, in determining the real seismic response of concrete building is highlighted. Comparison between three main structural systems for buildings is made and optimum use for each is identified.

Chapter 6

Reinforced Concrete Frame Systems

Abstract

Based on the theoretical background developed in the first five chapters of the book, concerning the behavior and analysis of reinforced concrete structures, the present chapter treats the main aspects of the analysis, design and detailing of frame structures of the earthquake prone concrete buildings. Analytical and laboratory investigations about the seismic behavior of concrete frames are synthetically presented and their impact on the optimum design and detailing is highlighted. Considering that a "well tailored" structure easily fulfills the structural requirements specific to seismic resistant structures, principles and rules for conceptual design of frames are defined. Modern analysis practice of reinforced concrete frame systems subjected to high intensity seismic actions, including advanced and simplified analysis under seismic equivalent forces, pushover and time-history post-elastic approach are presented. Advantages and disadvantages of these procedures are commented and their recommendable usage domain is indicated. The up-to-date design approach known as *capacity design method* is thoroughly presented and its application to the case of frames is discussed. Design and detailing recommendations for framed structural system in connection with specific seismic requirements and with the mutual interaction superstructure-infrastructure are commented. A case study about the seismic design and detailing of a concrete building framed section completes the chapter.

6.1 General Considerations

Frames are structures made of beams and columns rigidly connected in nodes (joints).

Reinforced concrete frames are natural extensions of stone, wood and (later) steel structures traditionally used as main skeleton for different types of buildings.

Cast-in-place reinforced concrete prismatic members naturally offer good flexural resistance and stiffness and are easily interconnected through rigid joints. Thus stiffening elements or braces are not necessary for reinforced concrete frames. We call them *moment resisting frames*.

The structural configuration and proportions, methods of analysis and design are strongly influenced by the building *seismic exposure*. According to this criterion, broadly, two distinct types of frames exist:

- Frames of buildings situated in non-seismic or very low seismic areas ("gravity-loads predominated frames") and
- Frames subjected to high intensity seismic actions.

Although the primary purpose of the non-seismic structures is to transfer *gravity loads* (dead and live loads), the lateral forces due to wind have to be taken into account. The taller a building is, the more significant will be, for its structural design, the lateral wind force. Consequently, the high-rise and super-high-rise frame systems located in non-seismic area have to observe principles and methods close to those of seismic resistant structures.

Even though the present book is focused on earthquake prone buildings, main features of framed systems under gravity loads combination will be briefly examined as a background for the study of frames subjected predominantly to seismic actions.

As already stated (see Chapter 5), for buildings subjected to high intensity seismic actions, the structural design process has to consider the whole structural system with its three components – superstructure, infrastructure and foundation soil. Accordingly, the present chapter deals, besides the frame superstructure (called also *structure*), with infrastructure too, taking into account the *mutual interaction* of three components of the structural system.

6.2 Behavior of Reinforced Concrete Frames

6.2.1 Behavior under Gravity Loads

Structural modeling and analysis of frame structures should take into account, as accurately as possible, their true mechanical behavior as well as the phenomena that occur during erection and progressive loading of the structure. Within current analysis and design procedures, some of these aspects are considered through relatively simple rules while others are ignored. However, for a correct understanding of the degree of accuracy of current analysis, most significant aspects of reinforced concrete frames behavior will be briefly discussed below.

(i) *Influence of construction history. Progressive loading of the structure* During the construction the building stories are progressively erected. Consequently, for each construction stage, a different load path acts on the already completed part of structure generating different distributions and magnitudes of the internal forces. This "construction history" is, generally, not taken into account within current structural analysis.

(ii) *Influence of differentiate cracking.* The structural members have sizes, reinforcement amounts and detailing according to the nature and maximum magnitude of the internal forces. At a certain threshold of external forces, cracks occur. The crack spacing, width, length and distribution along each element depend upon many factors including the reinforcement amount, number and sizes of reinforcing bars, internal force variation (along the element), etc. The cracks generate a substantial modification of sectional stiffness in comparison with un-cracked section. Accordingly, the internal forces distribution is modified. We call the redistribution of internal forces due to differentiate cracking *primary redistribution*.

(iii)   *Post-elastic force redistribution*. Accidentally, overloading (at least locally) of the structure can occur. Some critical sections, in such case, reach their capacity and post-elastic (plastic) phenomena are developed (yielding of longitudinal reinforcement accompanied by excessive cracking, potential local crushing of the compressed concrete, local anchorage loss, etc). Post-elastic deformations generate *post-elastic redistribution* of internal forces, called also *secondary redistribution*.

(iv)   *Time-dependent internal force redistribution*. Time-dependent phenomena like differential creep and shrinkage and foundations' settlements generate permanent internal force redistribution.

(v)   *Influence of factors not explicitly quantified by structural analysis*. It is difficult to list all the factors acting on a structure but which are not explicitly taken into account by the structural analysis. Such factors are the *imposed deformations* due to differentiate foundation settlement, temperature or shrinkage, geometrical errors during erecting process, etc.

The above considerations suggest the fact that the structural analysis has a strongly marked *conventional character*. We say that the structural design is performed with *nominal internal forces* which, sometimes, considerably differ from the real ones developed within the structure.

The differences between real and nominal forces are *globally* covered by the safety factors specific to each design approach and structural design code.

The designer has to use design value of loads according to the general design philosophy that involves the system of safety coefficients, methods of analysis, and rules of detailing consistent to each other. For all structures designed through the same approach and same set of codes, a uniform safety level is reached.

6.2.2   Seismic Behavior of Frames

Reinforced concrete frames subjected to seismic forces are among the most investigated structures. Numerous laboratory frame tests have been performed namely: (a) tests under lateral monotonic loadings (b) dynamic tests on shaking tables and (c) pseudo-dynamic tests (i.e. static tests under lateral loads simulating the seismic time-history loading). Besides the investigations on frame models (planar or spatial) much information was provided by testing frame components: columns, beams, beam-column joints or frame subsections. The spectacular development during last decades of computation facilities – both equipment and software packages – allowed performing extensive analytical research on frame seismic behavior using advanced numerical approaches. Precious lessons about structural seismic behavior have been provided by in-situ post-earthquake investigations. A thorough presentation of the up-to-date research on seismic behavior of frames can be found in the treatise Paulay and Priestley (1992). A good source of information is also the CEB State-of-the-Art Report (1996).

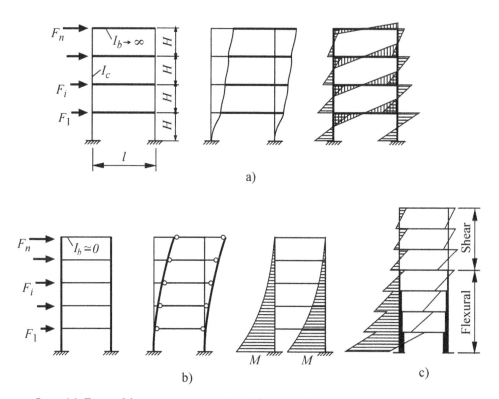

Figure 6.1 Types of frame response to lateral forces a) Shear-type response, b) Flexural-type response, c) Flexural & Shear response

Specific peculiarities of *elastic* seismic response of frames result from two main features: (a) stiffness ratio of frame components and (b) dynamic character of seismic action.

Fig. 6.1 shows typical bending moments under lateral forces for frames having different stiffness ratios between columns and beams: a) rigid beams and flexible columns, b) flexible beams and rigid columns and c) rigid columns at frame bottom and flexible column toward the frame top. For the a) frame, the behavior is dominated by the sliding of each floor over that of below. The flexible columns in respect to the beams (supposed to be "infinitely" rigid) are not effective in impeding this tendency so that the overall deformed shape of the frame is of shear-type. Accordingly, the structure can be called a *shear-type frame*. For the b) frame, the very flexible beams in respect to columns are not able to couple effectively the two columns which behave like two cantilevers having equal lateral displacements. Thus the two columns weakly connected through very flexible beams can be considered as a *flexural-type frame*. Actually these frames represent two limit cases. Depending upon the number of stories, upon the length of spans and magnitude of gravity loads, multi-story frames behave between these two limits. A frame like that of type c) in Figure 6.1 shows a shear-type behavior at its upper part since the bottom part (with strong columns) responds to the lateral forces through a flexural pattern.

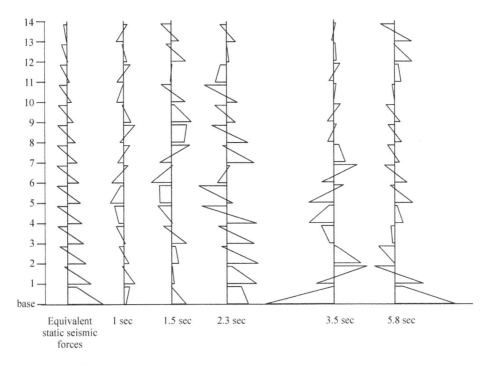

Figure 6.2 Time-history seismic response of a medium-rise frame column

Dynamic time-history response substantially differs from that under static mono-tonic loading. Figure 6.2 shows some typical sequences of time-history seismic response of a column of a medium-rise frame as compared with the bending moment dia-gram under static lateral loads. Predominance of different vibration modes in certain moments of the time-history seismic response is the source of those differences.

Post-elastic static and hysteretic behavior of reinforced concrete members has been examined within Chapter 2. Since frames are assemblies of such elements a first set of information about their post-elastic seismic response can be stated.

Depending upon the aspect ratio of frame components (beams or columns) or, more exactly, upon the magnitude of their shear arm M/Vh, the post-elastic response is dom-inated by bending or by shear; flexural-shear behavior can be also met (see Fig. 2.28).

Combining the information resulted from elastic static and dynamic time-history analysis with those related with post-elastic behavior of frame members overall features of frame response under severe seismic actions can be identified.

Post-elastic phenomena should be expected in frame components situated especially at the lower part of the structure. For medium- and high-rise buildings post-elastic phenomena occur also in zones situated toward 2/3 of the building height due to effect of higher vibration modes. The post-elastic deformations are more or less developed according to the degree of penetration into inelastic behavior domain. When the seismic response involves more reversal cycles specific crack patterns are recorded like X shape of shear cracks, anchorage loss of reinforcement, etc.

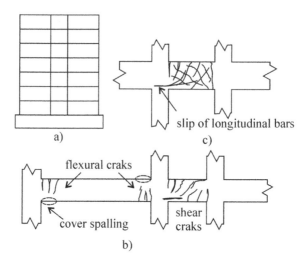

Figure 6.3 Crack patterns in frame beams a) Frame layout, b) Cracks in long and short beams, c) X-shape cracks due to reversal loading

It is important to point out that post-elastic phenomena means, actually, *structural damage* having different extents from very beginning of wide opened cracks due to initialization of reinforcement yielding up to local or generalized member failure.

Inelastic phenomena currently recorded in frame beams are of flexural, shear-flexural or shear type. Towards the ends of the long beams flexural plasticization is met characterized by: yielding of longitudinal reinforcement evidenced through large opened cracks, local crushing of concrete, spalling of concrete cover and longitudinal bar buckling. Short beams, like those of frames with a short central span, evidence shear-type damage (i.e. post-elastic phenomena) which can be identified through inclined (or X shape) cracks or cracks along the longitudinal bars due to their slip (Fig. 6.3).

The columns' post-elastic behavior is located at the frame bottom part since the internal forces – axial force due to gravity loads and to overturning moments, bending moment and shear force – have maximum magnitude in this zone. Most frequent post-elastic phenomena met in columns are: (1) compression failure accompanied currently by longitudinal bars plastic buckling, (2) plastic "hinges" (i.e. flexural post-elastic behavior shown by cracks perpendicular to element axe due to longitudinal bars yielding, local crushing of compressed concrete, cover spalling). They occur towards bottom and top ends of the long columns, (3) flexural-shear failure of medium length elements initiated through longitudinal bars yielding and completed through member split through an inclined critical crack, (4) brittle shear failure through a critical inclined crack.

Special attention should be paid to the column-beam joints. They are components which transfer high magnitude shear forces. Specific behavior, analysis and detailing of frame joints are examined in present chapter. Further information about behavior of framed structures during severe earthquakes is presented in Chapter 9.

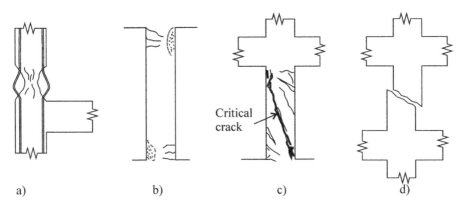

Critical
crack

a)   b)   c)   d)

Figure 6.4 Types of post-elastic phenomena or failure of frame columns a) Compression failure,
b) Plastic hinges, c) Flexural-shear failure, d) Shear failure

6.3 Analysis of Frame Structures

6.3.1 Advanced and Simplified Analysis under Seismic Equivalent Forces

Up-to-date structural analysis is, normally, computer-performed. Available specialized computer packages, having friendly interfaces, are very effective in determining
relevant vibration modes (periods and eigenvectors shapes), internal forces, displacements, stresses and strains, and any other parameters of structural response. They
easily allow modifying elements' sizes and general structural layout by adding or
deleting components, implementing different types of constitutive laws, structural
modeling, etc.

However, the misuse of facilities of computer-performed analysis easily can lead
to significant distortions to the correct quantification of real structural behavior. In
order to prevent that, the structural designer has to be aware of the limits of general
assumptions implemented into software and to react promptly to any "suspect" result
provided by the computer. He has to understand the global peculiarities of the expected
structural response, to be familiar with order of magnitude of response parameters, to
be able to perform alternate rapid evaluations of loads, response parameters and any
other relevant information. In respect with these requirements, brief presentation of
basic approximate analysis could be of interest. Still many code provisions in force are
based on simplified approach suitable for by-hand calculation.

Loads and load combinations

Generally, the frames are bearing elements for slab systems. It is accepted that slab
panels transfer to the bearing elements (secondary beams or frame girder) the loads that
act on *tributary areas* resulted by dividing the slab panel through 45° plans (Fig. 6.5).

It is accepted to consider an equivalent distributed load instead of triangular or
trapezoidal load distribution, which results from the shape of tributary areas of Figure
6.4a. The equivalence criterion is, generally, that of equal reactions.

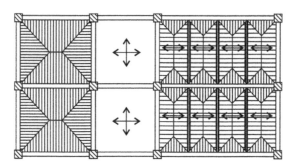

Figure 6.5 Tributary areas for gravity loads of frames

The *geometrical* model is obtained by replacing each prismatic element with its axis, i.e. the line that passes through the cross sections centroid. For rapid (approximate) analysis it is accepted to ignore the T shape of beams cross section, considering only its rectangular web.

Sectional stiffness design value has to take into account the following aspects:

- The frame beam cross section has a T-shape due to the interaction with the slab. However, the variation of bending moments along the beam modifies the effectiveness of flanges' contribution to strength and stiffness of the section. In zones where the flanges are compressed, the beam cross section effectively acts as a T shape while in zones with top side in tension the flanges are no more effective, being cracked.
- The sectional stiffness significantly differs for cracked and un-cracked sections. Accordingly, the sectional stiffness varies along each element.
- Further variation of the sectional stiffness along frame elements is due to other phenomena too, like: variation of the reinforcement amount, diagonal cracking due to shear, intensity and sign of axial forces, etc.

The conclusion is that, for reinforced concrete members, it is impossible to determine exactly the magnitude of sectional stiffness. Consequently, an average value of the sectional stiffness, constant along each element, is considered.

Generally the sectional flexural stiffness is expressed as functions of the gross moment of inertia I_c as for example:

- Beams: $(0.6–0.8)\, E_c I_c$ (E_c is the concrete elasticity modulus)
- Internal columns: $E_c I_c$
- External columns: $0.8 E_c I_c$.

Similar coefficients are provided by different design codes.

Computer structural modeling and analysis

Since up-to-date computer software packages for structural analysis are nowadays easily available and very effective it is recommended to use the most accurate models considering: spatial (3D) models, flexibility of the horizontal diaphragms and more relevant vibration modes. Interaction with the infrastructure and with the foundation soil has to be included in the analysis. The infrastructure connected to the frame super-structure comprises, besides the columns, structural walls, horizontal diaphragms and foundation components (footings or foundation beams or a general raft). Foundation soil can be considered as Winkler's elastic half-space modeled through elastic springs with appropriate (dynamic) stiffness.

Input data are implemented in the program through easy-to-use graphic interfaces. They comprise: geometrical and topological data (spans, bays, story heights, members' cross section, slabs' thickness and interconnection between elements), gravity perma-nent and live loads (elements' self-weight is, currently, automatically determined by the software), coefficients for determining sectional stiffness, data regarding seismic equivalent forces, different options concerning structural modeling.

Program performs linear elastic analysis determining eigenvalues and eigenvectors, modal internal forces and node displacements under horizontal seismic equiva-lent forces, modal effects combination and superposition of seismic action with gravity one.

Output results are presented in diagrams and/or in numerical format.

Current specialized software is provided with powerful facilities which allow mod-ifying the structural configuration (by adding or deleting elements) or properties of certain members. Accordingly, weaknesses of initial structure can be rectified and, finally, an optimal response is obtained which complies with code, functional and economic requirements.

Elastic analysis under equivalent forces is the basic tool for seismic structural design of buildings. Its results are enhanced through specific design procedures like the capacity design method and can be validated through post-elastic analysis.

6.3.2 Pushover Analysis

Specific features of post-elastic seismic behavior cannot be quantified through con-ventional elastic analysis with equivalent forces. Thereby, seismic performances of the building for different loading stages, location of plastic potential zones, magni-tude of plastic deformations, have to be more accurately examined using post-elastic procedures.

Pushover analysis is, in respect to these requirements, an attractive "low cost" tool, without excessive need of input data and processing time. Using advanced procedures, described in Chapter 4, pushover analysis is able to evidence the post-elastic behavior of planar and spatial frames, contribution of higher modes to structural response, premature brittle failure of different components, the effect of accidental failure of certain structural elements and other similar phenomena.

The results of pushover analysis are synthesized through *capacity curves* which provide an overall picture of the structural behavior which is easy to understand.

For regular buildings the use of planar models, according to principal axes, and conventional ("classical") approach with pointed plastic hinges offers a useful first set of information about structural behavior, complementary to the results of the elastic analysis.

It is recommended to use the *displacement controlled* version of the pushover analysis for identifying the weak structural components and the effect of their potential failure.

Besides general information about structural *virtual* behavior (i.e. under a supposed simplified loading – forces or imposed displacements) the post-elastic static analysis can be used for checking the basic requirement

Demand ≤ Capacity

in terms of *forces* or *displacements*. The most logical way to do that is to use post-elastic spectra for determining the structure's *seismic demands*. Displacement demand called *target displacement* is the post-elastic maximum lateral displacement at the building top level (or, alternatively, at a reference level) provided by the appropriate post-elastic spectra. It will be compared with the maximum capable lateral displacement provided by the pushover analysis.

Seismic force demand is the maximum total shear at the building base:

$$V_{\max} = S_{a,p} W_t / g \tag{6.1}$$

where $S_{a,p}$ is the post-elastic spectral acceleration and W_t – the weight of the total mass subjected to seismic action. It has to be compared with the lateral force capacity provided by the pushover analysis.

When the building seismic performances has to be checked, in more behavior stages according to the *seismic performance design approach,* advanced pushover analysis is recommended. It will quantify the nature and extend of damage (post-elastic deformations, damage of infill masonry, failure of certain elements, etc.) corresponding to each considered performance level (operational, life-safe, ultimate, etc.). Seismic demands (force or lateral displacement) will be quantified by the post-elastic spectra, corresponding to each design earthquake level.

6.3.3 Time-History Post-Elastic Analysis

It is commonly accepted that the time-history post-elastic analysis is the most powerful tool for simulating the "real" structural behavior under strong earthquake motions. However, the effectiveness of this approach strongly depends upon the accurateness of input data as follows:

(a) *Accelerogram.* It is presumed that the structural response to future earthquakes is similar or close to that determined through time-history analysis based on accelerograms recorded during past earthquake or artificially generated, which comply with local site conditions. In order to minimize the potential errors generated by this assumption the codes recommend resuming the analysis with at least 7 different accelerograms (recorded or artificially generated) and then to use either the average response value of all these analyses or the most unfavorable value.

(b) *Members' constitutive laws.* Models describing accurately the hysteretic post-elastic element behavior should be implemented in the analysis.

(c)   *Structural model.* Especially for structures with significant in-plan and/or in elevation irregularities spatial analysis would be required. Availability of such models is limited and, anyway, the amount of input data, storage space and processing time consumption are considerable.

Taking into account these limitations, it is obvious that the use of full 3D post-elastic time-history analysis is a tremendous task recommended only for very special cases. Otherwise simplified 2D time-history analysis according to building principal axes or elastic time-history approach can be used for obtaining approximate information about the dynamic (time dependent) seismic response. Because of the approximate character of such approach, the results have to be cautiously used.

6.4   Seismic Design of Frame Structures

The goal of seismic design is to determine the proportions of structural elements and their detailing so that the code requirements are fulfilled and the optimal configuration in respect with building function, easy construction and maintenance, labor and material economy, minimum need of post-seismic intervention, etc is observed. Accordingly, the simple application of codes is not enough. Besides general performance objectives and corresponding requirements provided by codes detailed analysis of each particular features of the future building have to be examined by the designers' team together with the investor.

6.4.1   *Preliminary Design*

From the very early stages of the design process reliable information about destination, general architectural layout and site peculiarities have to be provided to the structural designer. Besides the number of stories, in-plan and elevation building configuration it is important to know if it has a basement or not and, if yes, its scale (number of stories) and destination have to be specified. Information about soil characteristics and about the water table is crucial for selecting the solution for infrastructure.

The most used frame structures have the columns disposed according to the nodes of a rectangular regular mesh. Regularity in plan and elevation, uniform span and bay, advantageous building aspect ratio (total height/total width) are premises for obtaining through structural design a favorable, controlled seismic response.

Seismic Joints

In-plan building asymmetrical layout like T and L shapes generates very unfavorable torsional motions to the building which overload frames situated along the perimeter. Unfavorable response generates also building in-plan shapes like H or I with large flanges.

Irregularities in elevation (set-backs) are a source of sensitive zones due to the effect of high vibration modes and torsional effects.

When necessary, the unfavorable effects due to irregularities could be diminished or even eliminated by dividing the entire building through *seismic joints*. The joints' width is determined from the condition that building units separated by joints are dynamically independent to each other and the joint prevents pounding effects.

Number and location of seismic joints have to be carefully examined. One has to take into account that seismic joints could have a width as large as 15–30 cm or even more. Accordingly, the architectural and functional conditions are not easy to be fulfilled. On the other hand, one has to note that, under lateral seismic forces, the building sections behave like vertical cantilevers developing important overturning moments and shear forces. Transfer of these forces to the infrastructure and further to the foundation soil is a hard task. In respect to these issues, it is recommended to observe an aspect ratio (total height/total width) of each unit close to the optimal one. It is recommendable to observe an aspect ratio of the building (total height/total width) close to three; ratio up to five can be still accepted.

Infrastructure

Preliminary design has to deal with the whole structural system including infrastructure and foundation solution.

Because of cantilever behavior of the superstructure under lateral forces the foundation system has to be considered as a whole for the entire building.

Available solutions are listed and analyzed within the subchapter 6.7. However, some current solutions will be here briefly described.

Medium- and high-rise buildings are normally provided with a basement extended over one or more levels. Current use of basement is for parking but other functions can also be sheltered at this level. Basement always involves a peripheral wall, which is a structural component with high stiffness as compared with the superstructure. When some vertical walls at the basement level are required by the architectural needs they can be treated as structural walls interacting with horizontal slabs (over and within the basement). Peripheral wall interacting with internal walls together with floor slab systems (sometimes with raft, when necessary) acting as horizontal diaphragms constitutes a *multi-box system* extremely rigid and resistant which is the ideal infrastructure. Even for lower-rise buildings founded on poor soil a multi-box infrastructure is a solution which allows the building to "float" on the poor foundation soil avoiding expensive foundation solutions on piles.

Other issues to be taken into account during preliminary design

A structural designer has to keep in mind that, in contrast with gravity-dominated buildings, those subjected to high intensity seismic actions behave under lateral forces similar to a cantilever. Any architectural or structural solution which could generate brittle failures or any disadvantageous or uncontrolled effects should be avoided. Some examples are listed below.

Potential pounding between building sections or between adjacent buildings has to be prevented. Similar danger occurs when two neighboring buildings are connected through pedestrian bridges inappropriately treated.

Solutions which generate short columns or short beams with potential brittle shear failure have to be avoided. For example, frame with unequal spans should be preferred for a building with a longitudinal middle corridor instead of frames with a short midspan which lead to a short beam shear dominated (see Fig. 6.3).

Supplementary rules for obtaining good configuration should also be observed as much as possible:

- Try to obtain symmetrical plan shape
- Similar (close) stiffness of the whole system according to principal axes
- Uniform span and bay
- Limit torsional response
- Decouple elevator cages and staircases from the primary structure
- Prefer light-weight deformable partitions for preventing uncontrolled interaction with primary structure
- Keep uniform column cross section over the building height
- Avoid excessive building slenderness (recommendable total height/total width less than 3–5); this is an efficient way to ease the fulfillment of basic requirements

Steps of preliminary design

Preliminary design is an iterative trial-and-error process. Its goal is to "tailor" the structure so that the basic requirements are easily fulfilled. Thereby the proper structural design becomes a "fine tuning" process based on check up of all code requirements.

Best way to achieve a structural solution close to the optimal one is to use, from very early stages, good general computer software to determine the seismic response even though the input data are, initially, rough approximations. Such software allows performing easily modifications of initial model leading towards the desired optimal solution. Recommended steps are the following:

1. Based on architectural layout a structural system model is proposed with approximate sizes of all its components and estimated floor masses
2. Determine the response parameters and check if
   - First two modes are translations (not torsional type)
   - Vibration period T according two principal axes are close to each other
   - Equivalent total masses according first two (translational) modes differ less than 20%
   - Inter-story drifts over the whole building height comply with code requirements
3. Modify the structural model and resume step 2 until the above basic requirements are met.

6.4.2 Steps of Proper Design

Proper design is aimed to validate the sizes of all components of the structural system, to determine their reinforcement amount (flexural and transverse) and detailing. Criteria to be observed are those of reinforced concrete design rules, as stated by codes, as well as the fulfillment of specific requirements of seismic resistant structures.

As already stated (see Chapter 4), the seismic structural design is governed by the four basic requirements:

1. Strength and stability
2. Ensure favorable dissipating mechanism

3. Control of lateral displacements (or drifts)
4. Ensure local ductility of structure's members

At this stage of design, proper analysis models (currently 3D) have to be used involving superstructure, infrastructure and foundation soil. Permanent and live loads have to be accurately evaluated; load combinations according to code provisions will be specified as well as the stiffness of all structural components (beams, columns, slabs, infrastructure walls, soil).

Strength and stability is the first requirement to be examined even though, in many cases, the control of inter-story drifts and the capacity design method (corresponding to the second requirement) will lead to changes in members' sizes.

6.5 Capacity Design Method

Need for Plastic Deformations. Concept of Controlled Plastic Deformation

As already stated, in structures designed according to modern codes plastic deformations are expected under severe earthquakes. Acceptance of *post-elastic response* to high intensity seismic motions is the fundamental concept of up-to-date structural seismic design.

The *need* for development of plastic deformations during earthquakes is due to *economic reasons*. Design of structures for responding *elastically* to *any* seismic intensity involves resource consumption (financial and technical as well), unacceptable even for most wealthy societies.

The *possibility* to accept post-elastic structural and non-structural deformations results from the following:

- *Transitory* character of seismic actions (they have a duration of several seconds to several minutes).
- *Reversal* character of seismic inertial forces.
- *Low frequency* of high intensity earthquakes.
- Ability of reinforced concrete structures to develop *substantial post-elastic deformations* especially when certain specific measures are observed.
- *Reparability*, in most cases, of damage involved by development of post-elastic deformations.

The most significant progress of recent researches is, perhaps, the development of *methods* and *practical procedures* for *controlling* the seismic post-elastic response. This means that, nowadays, a *conception* and *design tools* are available for imposing, with high degree of credibility, *location of plastic deformation* (within the structure) and for limiting the extent of those deformations. The location of plastic deformation can be chosen in *naturally ductile* and *easy-to-repair* elements (Paulay, T., Bachmann, H. & Moser, K. 1990).

The *post-elastic controlled seismic response* is associated with the concept of *energy dissipating mechanism*.

The seismic energy input absorbed by structures is dissipated partially through damping and partially through the work of plastic moments (*"plastic energy dissipation"*), until the static equilibrium is reached.

The capacity of the whole structure to dissipate energy through plastic deformations depends upon the ductility of plastic zones and upon their number and location. When the post-elastic response of a structure involves a great number of plastic zones, located in elements with high ductility (like beams), the overall energy dissipating capacity of the structure is substantial. The structures that exhibit a limited number of plastic zones, located within elements with low ductility, possess a limited capacity to dissipate energy through plastic deformations and its ability to survive strong earthquakes is questionable.

Location of plastic zones within a structure depends upon the ratio between effective internal forces due to progressive loading and critical sections' resistance. If certain critical zones, considered to be suitable for developing plastic deformations, are appropriately designed and detailed while *all other zones* are endowed with enough capacity to resist elastically, the overall structure response to seismic motion will be *controlled* and not *random*.

Mechanisms for Plastic Energy Dissipation

The following discussion is more easily understood if we accept that the plastic deformations are concentrated in *plastic hinges* (see chapter 2).

Depending upon the stiffness of different frame members and resistance capacity of critical sections, different distributions of plastic hinges occur during strong earthquakes. The structure fails when enough plastic hinges occur to transform it in a kinematical *mechanism*. Different mechanism types can be encountered according to the plastic hinge distribution. Basically, there are two limit cases (Fig. 6.6):

(a) *Story mechanism* – the plastic hinges are located at the ends of columns of a single story ("soft and weak story") (Fig. 6.6a)
(b) *Beam mechanism* – the plastic hinges are located at the ends of all beams and at the columns' bottom (Fig. 6.6b).

The story mechanism involves a limited number plastic hinges located in columns, which are elements with low ductility (because of the compressive axial force). On the

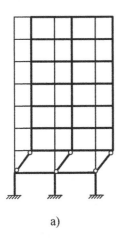 

a) b)

Figure 6.6 Types of dissipating mechanisms a) Story mechanism, b) Beam mechanism

other hand, the columns are *vital members* for the overall structure stability i.e. column failure can lead to building collapse. Since this mechanism involves a small number of plastic hinges, the amount of energy dissipated by each plastic hinge is substantial (total energy is shared within a limited number of hinges). Consequently, large plastic rotation demands are required for these hinges. It follows that the story mechanism is very unfavorable.

In contrast with story mechanism, the beam mechanism involves a great number of plastic hinges so that the amount of energy to be dissipated by each plastic hinge is small. Consequently, plastic rotations required within each plastic hinges are small so that the ductility requirement can be easily fulfilled. Moreover, the beams subjected to bending (without axial force) have naturally good ductile behavior without supplementary constructive measures. If a beam fails, the overall structure stability is not endangered.

The conclusion, which follows from above considerations, is that the beam mechanism is a favorable one. Accordingly, procedures should be defined and implemented in order to ensure a structural post-elastic behavior as close as possible to that model. For this purpose a practical method has been developed (Paulay, 1980) called *method of capacity design*.

Hinges that occur in beams rather than in columns characterize the favorable dissipating mechanism. Accordingly, that mechanism could be defined through the brief statement "weak beams-strong columns".

The method of capacity design fulfills the requirement of *weak beams-strong columns* by appropriately adapting the analysis based on equivalent seismic forces (current seismic design method). The basic judgment is the following: plastic deformations will develop in predicted sections if we design these sections to the internal forces resulted from conventional analysis while all other critical sections are proportioned and designed to the highest magnitude of internal forces which can be developed by the structure during earthquakes (Fig. 6.7). Consequently, all members of the structure, except the plastic zones, will behave elastically.

The overall plastic response of the structure will mobilize only imposed plastic zones, appropriately designed and detailed. According to this approach, the internal design forces within members supposed to behave elastically are considerably higher than those determined by the conventional analysis.

In other words, the structure should not be designed to provide "equal capacity" to all its components but an advantageous *capacities hierarchy* should be observed. Accordingly, chosen elements will meet earlier plastic deformations while others (having higher capacity) will respond elastically to the external input.

The key of the procedure is to assess, as accurately as possible, maximum capacities which can be developed by sections and elements. This means that, the actual resistance of a material, section or element can be substantially higher than that used for current design. We say that the structural components possess *over-strength* due to different factors. Three main over-strength factors can be identified:

(a) *Material over-strength* – the actual resistance of materials (steel, concrete, masonry) is normally higher that its design value

(b) *Over-strength due to constructive provisions* which oblige sometimes to appropriately increase the elements' sizes, reinforcement amount, etc. always on the safe side

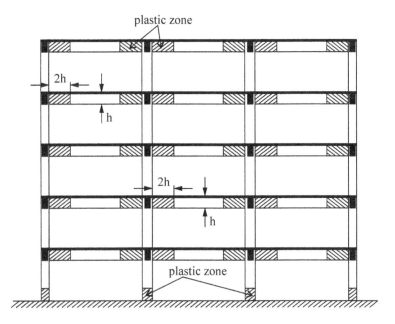

Figure 6.7 Potential plastic zones of a multistory frame

(c) *System over-strength* quantifies the capacity of a part of the structure or of the whole structure to develop higher resistance that required by design demand.

For example, for a reinforced concrete beam, part of a frame, we can define:

- Reinforcement over-strength factor λ_0 = actual average yield strength/design strength $f_{yd} \approx 1.25$
- Plastic hinge over-strength factor $\Phi_0 = \dfrac{M_0}{M_E} = \dfrac{A_s \lambda_0 f_{yd}(d - a/2)}{M_E}$

where: M_E is the bending moment resulted from elastic analysis under seismic forces, A_s – reinforcement actual area and other notation are shown in Figure 6.8.

Beam over-strength factor (i.e. system over-strength factor of sub-frame involving the continuous beam with n potential plastic hinges):

$$\psi_0 = \frac{\sum_1^n M_{o,i}}{\sum_1^n M_{E,i}} = \frac{\sum_1^n (\Phi_0 M_E)_i}{\sum_1^n M_{E,i}} \tag{6.2}$$

Steps of Capacity Design Method for Frames

The method of capacity design for frame structures involves the following steps:

1. Structural analysis under gravity loads and conventional (equivalent) seismic forces (elastic analysis). Internal forces (bending moments, shear forces, axial

forces) due to the two types of actions (gravity and seismic equivalent forces) are thus determined: M_G, V_G, N_G, respectively M_E, V_E, N_E.

2.   Choose potential plastic zones (desired). For frames, according to above considerations, plastic zones are expected to develop at beams' ends and at the columns' base sections.
3.   Design all potential plastic zones for *bending*, using the internal forces above determined. Detail all plastic zones for bending taking into account the rules for ensuring development of substantial plastic deformations without premature failure due to adverse phenomena (see subchapter 6.7).
4.   Calculate the flexural capacity of beam ends (plastic hinges) determined with the effective reinforcement amount, taking into account the over-strength of reinforcement (see Fig. 6.8).
5.   Determine the shear force in beams, associated to the flexural capacity of plastic zones determined with over-strength (see Fig. 6.10). This is the maximum shear force which can be developed within the beam.
6.   Design and detail of plastic zones for *shear* using conservative procedures provided by codes and the shear force above determined.
7.   Determine the *maximum magnitude of internal forces in columns*. For this purpose, calculate the magnitude of internal forces in columns associated with *moment capacity of beams with over-strength* and, thereafter, use appropriate correction factors for taking into account all other adverse phenomena that can be developed during earthquake: dynamic magnification of internal forces, effect of higher vibration modes, spatial effects (plastic hinges developed at all beam ends adjacent to the column).

Accordingly, the maximum internal forces in columns will be:

– Bending moments:

$$M = M_s^{col} \times \frac{\sum \Phi_0 M_r^{beams}}{\sum M_s^{beams}} \times K_M \qquad (6.3)$$

where K_M is a coefficient which take into account the adverse phenomena listed above.

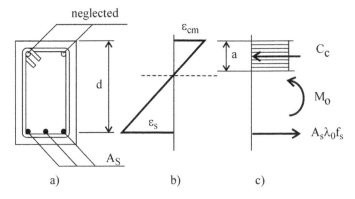

Figure 6.8   Bending resistance with over-strength a) Section, b) Strain, c) Stress

– Axial forces:

$$N = \sum V_{as} \tag{6.4}$$

8. Design and detailing of columns as common elastic elements.

Comments

1. It can be noted that, when capacity design method is implemented, the only internal forces determined through elastic (conventional) analysis directly used in design process (with their primary magnitudes) are the *bending moments in potential plastic zones of the structure*. The rest of internal forces initially determined are then adjusted (magnified) in order to prevent plasticization in undesired elements.

2. The capacity design method has a much wider range of use than the case of frames. Basically, it consists in adjustment of internal forces magnitudes, determined through conventional elastic analysis with equivalent seismic forces, so that the plastic deformations (which mean actually *structural damage*) are developed only in controlled easy-to-repair zones while the rest of the structure respond elastically to the seismic input. This objective is obtained through a suitable hierarchy of members' capacity. Although the procedure does not explicitly quantify the post-elastic behavior, it ensures a conservative control of post-elastic structure behavior. The advantage is that this task is accomplished through a simple approach, familiar to structural designers.

6.6 Drift Control of Frames Subjected to Seismic Actions

A multistory frame subjected to lateral forces shows a deformed shape like in Figure 6.9.

The total lateral displacement of a frame panel results from three independent factors (Fig. 6.9):

• Rigid body lateral displacement due to deformation of the structure below the considered level (Fig. 6.10b).

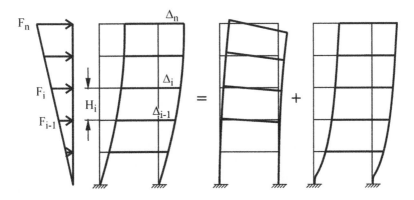

Figure 6.9 Deformed shape of a frame subjected to horizontal seismic forces

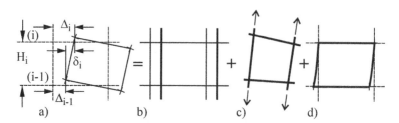

Figure 6.10 Deformation of a frame panel a) Deformation due to: b) Rigid body displacement, c) Columns' elongation and shortening, d) Frame panel distortion

- Overall deformation due to overturning moments. Overturning moments generate axial forces – tension, respectively compression – in frame columns. The columns' elongation, respectively shortening lead to the frame deformation shown in Figure 6.10c.
- Distortion of each frame panel due to the flexural deformation of its components (Fig. 6.10d). It can be quantified through the *inter-story relative displacement* $\delta_i = \Delta_i - \Delta_{i-1}$ or in dimensionless quantity of *inter-story drift* δ_i/H_i

The seismic displacement induces specific effects on overall building structural and non-structural response.

(a)   *Total lateral displacement* Δ (or *sway*) generates two main effects:
  - Creates an additional eccentricity of axial force in columns and, consequently, increases the bending moments ("second order effect"). This effect can be significant for slender columns, especially when post-elastic deformations occur.
  - The building occupants perceive it as a very uncomfortable, insecure, sensation that creates panic.
(b)   *Relative displacement* $\delta_i = \Delta_i - \Delta_{i-1}$ or *drift* δ_i/H_i induces damage to non-structural elements (infill masonry, partitions) which interact with the frames.

Taking into account the phenomena described above, related to the effects of lateral displacements on seismic behavior of buildings, the need for adequate *stiffness for lateral forces* becomes evident.

The stiffness shall be effective for *total lateral displacement* as well as for *inter-story drift* at all frame levels.

This is a basic requirement for structures subjected to high intensity seismic motions.

In order to fulfill correctly this requirement, answers to the following questions are needed:

(a)   Which level (or levels) of seismic action has to be considered?
(b)   Which are the acceptable magnitudes of lateral displacements and drifts that may be allowed for reasonably limiting their effects?
(c)   How the lateral displacements should be determined for checking-up reliably the fulfillment of this requirement?

Nowadays only partial and transitional answers can be given to these questions. The seismic design codes of different countries show a great range of provisions corresponding to this requirement. Significant progress in this field is expected once the *performance-based seismic design* (see sub-chapter 4.9) will become effective.

When the drift control refers to the ultimate limit state, the following procedures for determining ultimate displacements (i.e. displacement demand) could be used:

- Multiplying the displacements calculated for elastic structure under equivalent seismic forces by q (rule "equal displacements" was accepted) (see Sub-chapter 4.7 and Figure 4.10)
- Determine displacement demand from inelastic spectra
- Performing a post-elastic dynamic analysis using the design accelerograms that complies with the site conditions. The maximum displacements resulted from this analysis will be used to determine maximum drifts.

6.7   Local Ductility of Frame Components

Within frame members with expected post-elastic deformations (having potential plastic zones), specific provisions have to be implemented in order to ensure their ductile behavior. In other words, potential brittle failure of these zones has to be prevented.

Beams

The frame beams are subjected to bending and shear force.

Normally, under *bending moments* a ductile behavior is naturally ensured, while *shear force* induces a brittle failure.

(a)   *Flexural ductility*
Although the flexural behavior of beams is generally ductile, some rules have to be observed.

Potential plastic zones of beams have to be reinforced with *ductile steel* only. The use of cold work steel in potential plastic zones has to be avoided.

Moderate longitudinal steel percentage has to be used (the recommended range is between 0.8 and 1.2% in beams).

A phenomenon which can lead to a premature lost of ductility is *the post-elastic buckling* of the compressed steel. Due to alternate loading, both longitudinal reinforcement layers (at the top and at the bottom of the beam) can be subjected to compression. Since post-elastic deformations are accepted within potential plastic zones of the beam, the *post-elastic buckling* of the compressed steel can occur, with the corresponding damage of the surrounding concrete. In order to prevent this buckling, *stabilization stirrups* spaced at less $(4-6)\phi_{min}$ (ϕ_{min} is minimum bar diameter of the section) have to be provided. This means that stirrups in potential plastic zones have two functions:
– Stabilization of compressed bars (to prevent buckling);
– Resist shearing forces (these stirrups will be calculated).

(b)   *Effect of the shear force*
The ductile behavior of beams is effective when large flexural deformations (rotations) can be developed. So, the premature brittle failure through shear has to

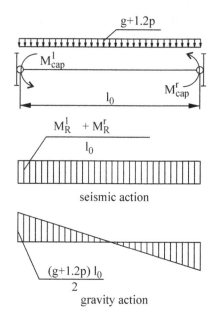

Figure 6.11 Shear force associated to the yielding of beam-ends

be prevented in order to allow developing substantial plastic flexural deformations. In other words, an appropriate *over-resistance to* shear (in comparison with bending) has to be observed.

This requirement is achieved by two complementary measures:

- Designing transverse reinforcement for maximum magnitude of shear force which can be developed within the plastic zone
- Using conservative procedures to design for shear.

The *maximum magnitude of shear force*, which can be developed in a beam, is related to the maximum bending capacity of the beam-ends. It is accepted that, at both beam-ends, plastic hinges are developed. Thus, the shear force associated to the bending moment capacity of plastic hinges, will be (Fig. 6.11):

$$V_{as} = \Phi_0 \frac{|M_R^l| + |M_R^r|}{l_0} + \frac{(g + 1.2p)l_0}{2} \tag{6.5}$$

where: M_R^l and M_R^r are the flexural capacities of the left end, respectively right end sections of the member, determined with the real reinforcement area provided within the sections, taking into account the *potential over strength of the reinforcement* $(f_y^* = \lambda_0 f_{yd} = 1.35 f_{yd})$
Φ_0 – sectional over strength factor and
l_0 – spacing between plastic hinges.

The *material* over strength (for reinforcing steel) is due to potential yielding limit increase through strain hardening and to statistical potential difference between design and average strength. The material over strength is propagated to sectional level generating a *sectional capacity over strength*.

The shear resistance of the beam is ensured by the stirrups and by the shear resistance of compressed concrete:

$$V_R = V_{stirr} + V_{concr} \tag{6.6}$$

The concrete within plastic zones can be damaged due to cyclic loading. Thus, a conservative shear capacity of the beam shall be determined accepting a diminished contribution of the concrete (in comparison with that corresponding to gravity loads).

The stirrups area and spacing will be determined from the inequality:

$$V_{as} \leq V_{Rs} \tag{6.7}$$

Columns

For columns, elastic behavior is expected if the capacity method is properly applied. Accordingly, no special provisions should be observed for columns, excepting the potential plastic zones.

Within potential column plastic zones, ductile behavior shall be ensured.

Columns are subjected to bending moments and axial forces $(M + N)$ and to shear force.

The behavior for *eccentric compression* $(M + N)$ is governed by the magnitude of the dimensionless axial force (see Paragraph 2.3.3):

$$n = \frac{N}{A_c f_c} \tag{6.8}$$

According to the magnitude of this parameter three behavior types until failure have been evidenced:

(a)   For $n = 0.0$–0.2 – ductile behavior similar to that of elements subjected to pure bending
(b)   For $n = 0.3$–0.5 – semi-ductile behavior; failure is caused by the compressed concrete crushing (ultimate concrete strain is exceeded)
(c)   For $n > 0.5$ – elastic-brittle behavior

Consequently, in order to ensure ductile behavior of column plastic potential zone, for each of above cases the following measures have to be taken:

(a)   For $n = 0.0$–0.2 – no special measures are required except those related with preventing the buckling of longitudinal reinforcement, i.e. stabilizing stirrups (spaced at less than 6 d_{min})
(b)   For $n = 0.3$–0.5 – the ductility is improved if the concrete is confined so that it develops large limit strains. Thus, within potential plastic zones of columns, with $n = 0.3$–0.5, hoops or spiral transverse reinforcement have to be provided.

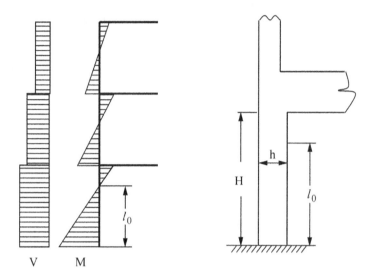

Figure 6.12 Shear arm and aspect ratio of the columns

(c) $n > 0.5$ has to be avoided for columns with potential plastic zones, by increasing their cross section.

Within potential plastic zone no lapped splices are allowed. Accordingly, the connection of longitudinal bars has to be made outside of the potential plastic zone or special connectors have to be used.

The shear behavior of the columns is governed by the magnitude of the *shear arm*:

$$a_{sh} = \frac{M}{Vh} \tag{6.9}$$

Since throughout the column height the bending moment varies linearly and the shear force is constant, the dimensionless magnitude of the shear arm a_{sh}/h (h is the cross section height) is equivalent to the *aspect ratio*:

$$\lambda = \frac{l_0}{h}$$

where l_0 is the distance between the section with maximum moment and that with zero moment (Fig. 6.12). (For very rigid beams in respect with columns $l_0 = H/2$).

According to the aspect ratio, the shear behavior of the columns evidences three distinct patterns (see Fig. 2.28):

- For *short columns* ($\lambda < \approx 2.5$) the failure is governed by shear. It has a brittle character and occurs through a critical inclined crack. No effective way exists to ensure a ductile failure of these elements so that no post-elastic deformations are allowed in short columns.

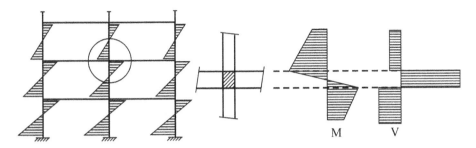

Figure 6.13 Shear forces due to seismic action within beam-column joints

- *Middle-length columns* (λ = about 2.5 to 5) show a semi-ductile flexure-shear failure. This means that, for these columns, the failure starts by the yielding of longitudinal reinforcement within the zones with maximum bending moments and is completed by a shear inclined critical crack. If the premature failure due to shear is prevented, the column is able to develop substantial (flexural) deformations. Accordingly, stirrups (size and spacing) have to be calculated with the formula used for the beams, for the shear force associated with the plastic moments at the column ends. The stirrups will be extended over the whole column height.
- *Long columns* (λ > about 5) are predominantly subjected to bending (with axial force) and, consequently, have a "natural" ductile behavior, with plastic hinges at both ends.

Of course, this implies that the column slenderness is not critical from the point of view of the buckling.

6.8 Beam-Column Joints

The importance of beam-column joints in seismic behavior of frames was recognized only in late the late 1980s.

The substantial progress during the last decades in understanding the structural behavior under sever seismic actions led to a gradual improving of design and detailing of frame beams and columns. They became more and more strong, provided with appropriate reinforcement, both longitudinal and transverse. In the absence of appropriate design and detailing measures, the beam-column joint became the "weak link" of the "structural chain", being responsible, in certain cases, for the failure of reinforced concrete frame buildings.

Within beam-column joints complex stress – strain states are developed due to its specific proportions, to the internal forces transferred by the adjacent elements and by the reversal character of the seismic action.

The most important features of joints behavior are demonstrated by the presence of high magnitude shear forces. This can be explained by the high moment gradient within the joint core (Fig. 6.13).

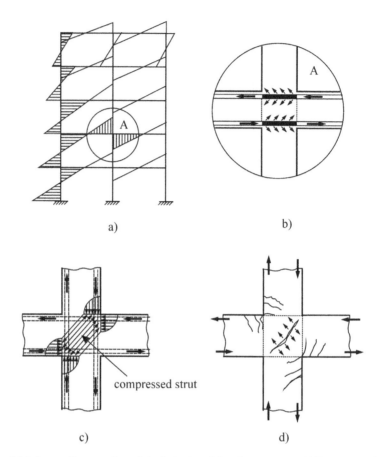

Figure 6.14 Internal beam-column joint behavior a) Bending moments, b) Forces on rebar crossing the joint, c) Internal forces in joint, d) Joint cracks

Another significant effect is related to the forces that act on the beam longitudinal reinforcement under predominant seismic action. Under this action, the bending moment changes its sign in the sections adjacent to the joint (Fig. 6.14).

Thus, the longitudinal beam rebar is subjected to compression at one side of the joint and to tension at another side and tends to be pulling out from the joint. Accordingly, the rebar has to transfer substantial sliding forces, through bond stresses, within the joint. Moreover, the sense of these forces changes, due to reversal seismic forces, which creates particularly severe conditions. The phenomenon is worsened when plastic deformation of the steel penetrates within the joint.

If the steel looses its bond, the joint is no longer able to transfer the forces of adjacent beams. So, the joint stiffness and, consequently, the coupling effect of the beam are drastically diminished, which is extremely dangerous for the frame resistance capacity.

On the other hand, the post-earthquake repair of a damaged joint is extremely difficult (practically impossible).

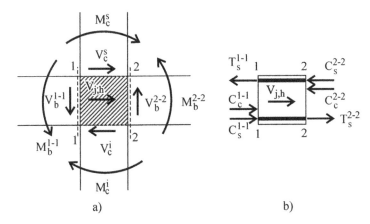

Figure 6.15 Horizontal shear force in beam-column joint a) Internal forces, b) Free body equilibrium

Taking into account these behavior peculiarities and considerable difficulties to repair damaged joints after severe earthquakes, appropriate measures for assuring joints *elastic behavior* and to prevent any unfavorable phenomena above described are recommended.

The basic requirements for beam-column joints design and detailing are, thus, the following:

1. To ensure the *resistance capacity,* under most unfavorable force combination in any behavior stage
2. To keep significantly unaltered the *stiffness* under internal forces associated with yielding of adjacent members when subjected to cyclic actions
3. To ensure bond and anchorage of the beam longitudinal bars crossing the joint.

Shear Forces within Beam-Column Joints

Most significant internal forces for beam-column joints are the horizontal and the vertical *shear force* $V_{j,h}$ and $V_{j,v}$.

Their magnitude can be derived from the equilibrium of the free body shown in Figure 6.15.

The shear force within joint is (Fig. 6.15b):

$$V_{j,h} = T_s^{1-1} + C_s^{2-2} + C_c^{2-2} - V_c^s \tag{6.10}$$

But, within the cross section 2-2, the compressive force $C_s^{2-2} + C_c^{2-2}$ is balanced by the total tensile force of bottom reinforcement:

$$C_s^{2-2} + C_c^{2-2} = T_s^{2-2} \tag{6.11}$$

and, so, the horizontal shear force will be:

$$V_{j,h} = T_s^{1-1} + T_s^{2-2} - V_c^s \tag{6.12}$$

Accepting the conservative assumption according to which, at both sections 1-1 and 2-2 adjacent to the joint, plastic hinges occur, the horizontal shear force at joint results:

$$V_{j,b} = (A_s^{1\text{-}1} + A_s^{2\text{-}2})f_{s,y} - V_c^s \qquad (6.13)$$

($f_{s,y}$ is the steel yielding limit).

Since tangential stresses, considered uniformly distributed throughout the joint horizontal and vertical plans, are equal (principle of tangential stress duality), the vertical shear force within the joint is:

$$V_{j,v} = \frac{h_b}{h_c} V_{j,b} \qquad (6.14)$$

Over the joint height horizontal stirrups have to be provided, dimensioned to resist the shear force $V_{j,b}$. Their diameter and spacing will be not less than those of column's stirrups.

To resist vertical shear force $V_{j,v}$, instead of vertical stirrups, it is more feasible to extend the column vertical reinforcement over the joint height. At least three bars have to be provided on each joint face.

More detailed rules for detailing of joints are provided by codes. A thorough examination of behavior, design and detailing of beam-column connection can be found in the treatise of Paulay and Priestley (1992).

6.9   Interaction Frames/Masonry Infill

For functional purposes, frame buildings are often provided with infill *walls* or *partitions*. Clay or light-concrete brick masonry, light porous concrete blocks or super-light materials like gypsum-cardboard panels are used for constructing these infill panels.

Since the masonry infill weight is supported by the principal frame structure, they are generally considered as *non-structural members* and their interaction is ignored in most cases.

However, interaction between frames and masonry infill can modify, sometimes considerably, the expected seismic response of the structure. Two limit situations can be encountered:

(a) *Relative strong masonry infill in respect to frame stiffness and resistance.* During earthquakes, the masonry infill is effective in impeding the free lateral sway of the frame. In the early stages of deformation the masonry behaves elastically and the stiffness of composite structure (in-filled frame) is considerably increased. Due to interaction with masonry, unexpected forces act against frame components (beam or column) which can be seriously damaged, sometimes in a brittle manner (shear failure as an effect of "short column" or "short beam").

(b) *Weak infill in respect with the frame.* Distortion of the framework creates a "vice effect" clamping diagonally the masonry panels, which crack, fail or fall down.

Intermediary situation between these two limits frequently occur generating damage of differing severity to both infill panels and frames. Nevertheless, in some cases, the presence of infill masonry within frame structures, inadequately designed for seismic

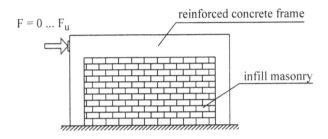

Figure 6.16 Experimental specimen for investigating frame/wall interaction

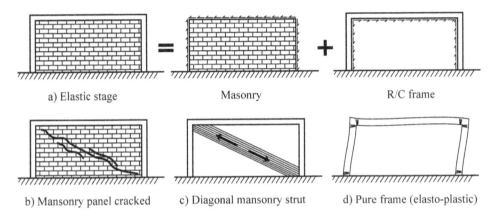

Figure 6.17 Stages of frame-infill interaction

actions, led to unexpected over strength of the overall structure ensuring the building survival.

Consequently, considerable effort has been made – both experimental and analytical – to understand the specific mechanisms of that interaction, to define adequate analytical models and to deduce design and detailing appropriate rules.

Typical experimental specimen for investigation of the frame/infill wall interaction is a frame panel filled-in with brick masonry (Fig. 6.16).

The fill-in wall can be provided with opening of different sizes.

The in-plane behavior of such specimens is investigated by loading with monotonic horizontal force or with imposed horizontal displacement.

The loading up to the failure evidences the following *behavior stages*:

1. *Fully composite elastic action* – for low magnitude of external loading both frame and wall behave elastically. If a full contact between two components exists, the frame transfers to the wall a fraction of the external force through shear boundary stresses (Fig. 6.17).

    The wall panel is subjected to shear. Its primary capacity can be reached through
    • *Sliding along horizontal mortar course* – this is the case of masonry with low strength mortar

- *Diagonal cracks* due to principal tensile stresses, typical for masonry with relatively good strength mortars.

Although a critical stage for masonry wall is reached, its effective failure does not occur due to the clamping effect of boundary elements. Instead, the continuous contact between wall and frame is broken along certain lengths and a masonry compressed strut is formed.

2. *Bracing action.* The masonry compressed strut tends to impede the free flexural deformation of the frame. Consequently, the composite structure (frame/infill) behaves similar to a braced frame with diagonal compressed struts.

In comparison with the original pure flexural frame the braced structures possesses considerable higher *stiffness* and *resistance.*

So, the bracing action of the infill walls enhances (in many cases substantially) the capacity of the primary structure to resist earthquakes. This is the explanation for unexpectedly good seismic performances of some multistory structures that have been designed only for gravity loads but survived sever earthquakes.

The bracing action is valid provided that the compressive stresses of the strut are less than masonry strength. When the strut strength is reached, the wall effectively falls down and only the frame carries off the external load.

3. *Pure frame action.* Since no more infill wall exists, the structure acts for further increases of external load as *pure (flexural) elastic-plastic frame.*

The above-described behavior is effective provided that

- Good, continuous contact exists between wall panel and concrete boundary elements;
- Failure is governed by wall strength and not by frame overall resistance ("weak wall-strong frame").

When these assumptions are not met, other failure modes occur and the presence of fill-in wall can worsen the frame seismic performances. Some such examples will be briefly described below.

(a) *No wall-beam contact.* It is common that the upper course of the wall panel (below the beam) is not properly filled with mortar. In such cases, the compressed wall strut is pushing against the frame column (Fig. 6.18). So, the column is subjected to a transverse force generating additional bending moments and, especially, high shear forces. Generally, a "short-column effect" occurs with specific damage (brittle shear failure).

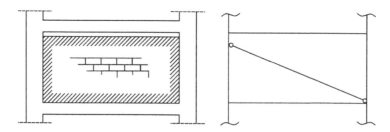

Figure 6.18 Infill panel with no contact with the beam

(b) *Masonry infill partially extended over the story height.* To accommodate windows in façades, sometimes the masonry wall is extended only partially over the story height (Fig. 6.19). If the masonry has a good strength, the frame columns become *short columns*, with considerable higher stiffness, absorbing, consequently, higher lateral forces but showing a poor, brittle shear behavior.

(c) *Masonry panel with openings.* When door openings are provided near the columns, the compressed strut tends to push against the beam generating internal forces opposite to those due to gravity loads (Fig. 6.20). Unexpected damage and failure can occur in these cases.

The conclusions of above considerations can be summarized as follows:

• Interaction between frames and infill walls generates effects generally not explicitly taken into account during the seismic design process.
• Although a potentially favorable effect can be generated by this interaction, leading to better seismic performances (evident, especially, for frames not designed for seismic actions), damage of walls and partitions shall be expected during strong earthquakes. Sometimes, the interaction with infill causes additional damage to frames (in comparison with frames free of walls).
• Extent of damage depends upon the magnitude of relative *inter-story drift*. i.e. ratio of relative displacement versus story height. This can be considered as an index of *potential damage* of partitions and fill-in walls and of structural elements as well, due to frame/walls interaction.

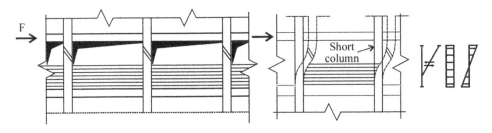

Figure 6.19 Masonry infill partially extended over the story height

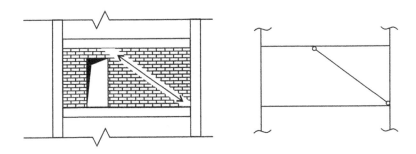

Figure 6.20 Infill masonry panels with openings

For quantifying the interaction frame/infill masonry walls excessive sophisticated models are worthless since the brick masonry and its interface with frame show pronounced uncertainties of mechanical properties.

A reasonable model for masonry infill is with compressed struts.

The average width of equivalent strut w can be taken as much as (Paulay & Priestley, 1992):

$$w = 0.25d_m \tag{6.15}$$

where d_m is the strut length. For its elasticity modulus E_m same authors recommend the value:

$$E_m = 750f'_m.$$

Note that the bracing action of infill masonry modeled through compressed struts is effective for stage (e) in Figure 6.16, i.e. after partially detaching of infill from frame (tensile masonry strength has been reached).

Performing a pushover analysis of composite structure (frame with masonry infill) the seismic performance defined as "damage of nonstructural elements" is identified (see Fig. 4.12).

6.10 Infrastructures and Foundations

6.10.1 *General*

Foundations are structural components that transfer the internal forces of the structure to the soil.

The load transfer path considerably differs for frames subjected predominantly to gravity loads and for those subjected to high intensity seismic actions.

For *gravity-predominated frames*, the external loads act, primarily, on slabs. The slabs and their supporting elements – beams – transfer the gravity loads toward the columns. Further, the columns transfer the loads toward their bearing elements supported by the soil, mainly by *axial forces* (bending moments and corresponding shear forces are of limited magnitude). Consequently, each column support – the foundation – can be studied as individual element. Thus, generally, the foundations of *gravity-predominated frames* are provided for each individual column as isolated footing. Some particular cases will be discussed below.

The buildings subjected to major earthquakes act as vertical cantilever under lateral seismic forces, developing globally important *overturning moments* and *shear forces*. Accordingly, the whole structure has to be provided with a global foundation system aimed to transfer to the soil the seismic overturning moments and shear forces, resulted from the cantilever effect, as well as the gravity loads. We call this foundation system *infrastructure*. In order to be effective support for superstructure the infrastructure has to have substantially higher resistance and lateral stiffness.

Depending upon the number of stories (total building height), seismic zone intensity, plane shape of the building and other relevant parameters different solutions for the infrastructure can be adopted.

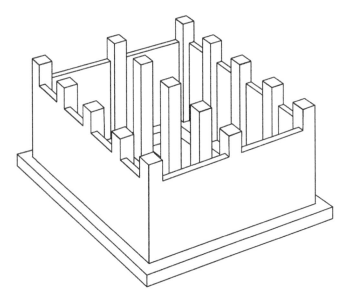

Figure 6.21 Peripheral retaining wall of a building basement

Generally, the high stiffness and resistance of the infrastructure is obtained by a system of additional structural walls provided at basement level.

The most frequent solutions for framed building infrastructures are examined below.

6.10.2 *Infrastructure with Peripheral Walls*

Multi-story buildings are often provided with basements used for different purposes: garage and parking, technical facilities, archive, etc. The basement has to be separated from surrounding soil by a *peripheral wall* which acts as a retaining wall. Peripheral walls interact with the slab over basement and with their own foundations forming a rigid box. They are provided with continuous footings that ensure developing appropriate pressure on the soil. For low-rise buildings, this box is generally able to transfer to the soil the seismic and gravity loads of the superstructure acting as an *infrastructure*.

The internal columns are kept continuous over the basement height and are provided with normal (individual) footings.

The total horizontal seismic force associated with yielding mechanism (determined as shown in Fig. 6.22) is carried by the two walls parallel to the force (Fig. 6.23).

The total horizontal force at the interface superstructure/infrastructure is collected and, then, transferred to the walls by the horizontal diaphragm which is the slab over basement. Thus, the slab over basement has a double function:

1. It is a component of the slab-beam system, transferring vertical forces toward its supports (peripheral walls, beams and columns);
2. Acts as a horizontal diaphragm, transferring the horizontal forces to the vertical walls.

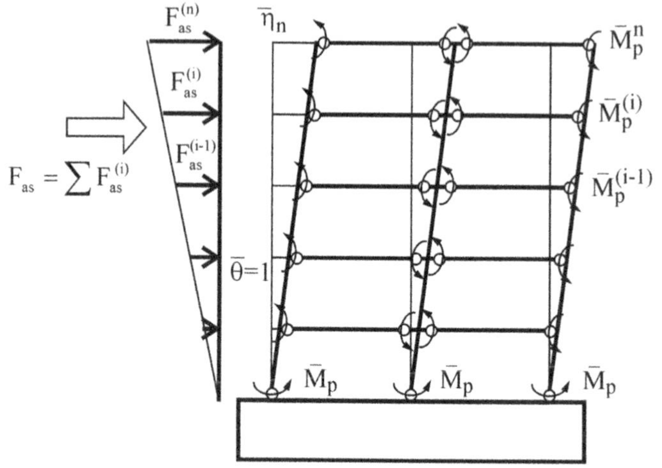

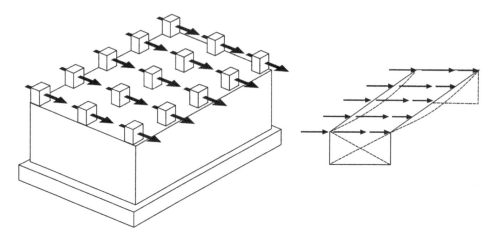

Figure 6.22 Use of principle of virtual work for determining the seismic forces associated to the yielding mechanism

Figure 6.23 Transfer of seismic forces to the infrastructure walls

According to its *diaphragm function,* the slab acts like a horizontal beam supported by the walls (Fig. 6.24) Thus, the diaphragm behaves like a beam spanning over the length of the building loaded with shear forces associated to plastic moments calculated with over strength at the base of all columns of the superstructure (Fig. 6.22).

Analysis and *design* of foundation system are performed assuming that, under design seismic action, infrastructure components as well as the foundation soil behave elastically. Thus, principles of capacity design method (see 6.8), are applied. Accordingly, the infrastructure's loading results from the plastic moments at the bottom of the columns, determined with over-strength, and their associated axial and shear forces.

It is supposed that the superstructure has been dimensioned and detailed according to the rule *weak beams-strong columns* (capacity design method).

At the interface superstructure/infrastructure acts the total horizontal seismic force associated to the plastic moments within upper-structure calculated with over-strength ("seismic capacity" or "seismic resistance" of the superstructure determined with plastic moment's over-strength) F_{as} and the total overturning moment generated by this horizontal force. According to a simplified, conservative approach the superstructure seismic capacity can be easily determined using the *principle of virtual work*. It is assumed that the failure mechanism of the superstructure is reached and within plastic hinges act the plastic moments determined with over-strength:

$$\overline{M}_p = \Phi M_p$$

where $\overline{M}_p$ is the plastic moment with over-strength, Φ – the sectional over-strength factor (1.25 to 1.35 depending upon codes) and M_p – flexural capacity ("plastic moment") of plastic hinges.

Vertical axial forces of external (peripheral) and internal columns, generated by gravity loads and by the seismic action corresponding to the plastic mechanism, have to be added. The peripheral walls are subjected also to the soil pressure which has to be taken into account.

With these forces computer analysis can be performed modeling the walls and the horizontal diaphragm (slab over basement) with shell-type finite elements and the soil with Winkler's springs.

An alternative simplified procedure consists in accepting that the internal forces and stresses within infrastructure components determined initially through a global elastic analysis of the structural system (superstructure & infrastructure & foundation soil) *multiplied to* Φ are close to those which ensure elastic response to the infrastructure.

Design and Detailing

Infrastructure components (walls, diaphragm, columns, and footings) will be designed and detailed according to their internal forces determined by the analysis and to specific code provisions. Note that the walls are of *squat* type, shear force predominated (see also Paulay & Priestley, 1992).

Safe link between diaphragm and walls have to be ensured. The interface between these elements is, normally, a casting joint that shall be accurately executed. It has to be crossed by connectors designed through the *shear friction procedure* (Fig. 6.24). Accordingly, the total area of the connectors along the sliding plane (between diaphragm and foundation) will be determined from the inequality:

$$V < V_R \tag{6.16}$$

where: V is the total shear force along the joint.

Critical situations can occur if the slab is provided with large openings. In such cases, the horizontal diaphragm behaves like a frame having rigid zones connected with weak links (portion of the slab adjacent to the opening). It could be possible that supplementary slab reinforcement is required along the opening contour. Another potential critical situation is that of the prefabricated slabs. Normally, concreting of a thickness of 40–60 mm reinforced with welded wire fabric made meshes will be provided over the prefabricated slab.

This approach is simplified but on the safe side. A more accurate analysis would consider the infrastructure as an elastic box supported by elastic continuum (i.e. the

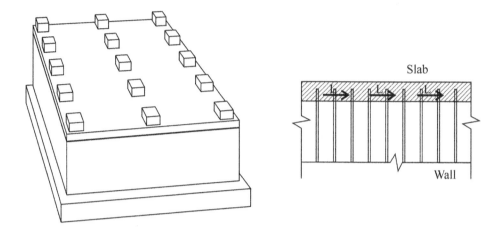

Figure 6.24 Connectors between infrastructure wall and horizontal diaphragm

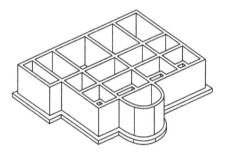

Figure 6.25 Infrastructure with multiple internal walls

foundation soil). The forces acting on this box are the internal forces at each column base (plastic moments determined with over-strength and associated axial and shear forces). The stresses and the total internal forces for each wall as well as more accurate distribution of the soil pressure can be determined by using finite element software.

6.10.3 Infrastructure Made by a System of Walls and Diaphragms

Medium and high-rise buildings subjected to high intensity seismic actions require strong infrastructures. In such cases more intermediary structural walls have to be provided at the infrastructure level. When a multi-level basement is required, for architectural and functional needs, the infrastructure consists of more vertical structural walls (along the basement perimeter as well as intermediary walls with or without openings) interconnected with horizontal diaphragms (slabs). If the foundation soil is weak or in the case of high-rise buildings in high-intensity seismic zones the walls' foundation is extended over the whole building area, forming a *raft* (thick slab). The walls system interacting with the raft and with the horizontal diaphragms creates a *closed multi-box system*, extended over the whole basement, showing a high torsional and flexural stiffness (Fig. 6.25 and 6.26).

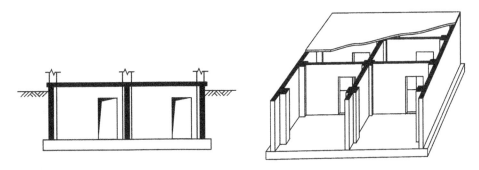

Figure 6.26 Infrastructure with closed multi-box system

Normally, multi-box systems should be calculated with a finite element program, taking into account the interaction with the soil modeled with Winkler elastic springs or as an elastic continuous 3D solid. The infrastructure is loaded with the flexural capacity of the columns' base determined with over-strength and their associated shear and normal forces.

6.11 Case Study

6.11.1 *General Layout. Input data*

The case study deals with a 9-story and one level basement office building. For obtaining large area open spaces the structure was conceived with frame structure. The basement accommodates archives, technical equipment, and other similar functions. The building consists of more identical sections separated by seismic joints. Each section has a total area of $520 \, m^2$.

Building is located in a high intensity seismic zone having design peak ground acceleration (PGA) with a return period of 100 years corresponding to the no-collapse requirement ("ultimate limit state ULS"): $a_g = 0.24 \, g$. For service limit state (SLS) the design earthquake return period is considered 30 years and the corresponding PGA is

$$a_{gs} = 0.50 \, a_g = 0.12 \, g \tag{6.17}$$

Admissible drift: 0.008 for SLS and 0.025 for ULS.

Due to important seismic exposure (quantified through high peak ground accelerations) and to total building height, the structure is *seismic dominated*.

Loadings (characteristic magnitudes):

1. Permanent and quasi-permanent actions (besides the self-weight of primary structure members): $3.75 \, (kN/m^2)$
2. Variable actions (for office function): $2.0 \, (kN/m^2)$
3. Snow (on roof): $2.0 \, (kN/m^2)$

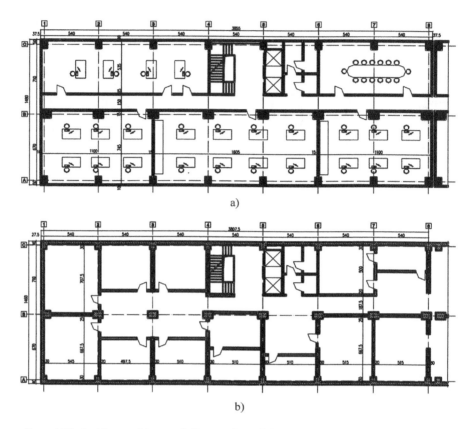

Figure 6.27 Architectural layout: a) Current floor, b) Basement

According to different combinations to be taken into account appropriate factors will multiply the magnitude of each action.

These nominally distributed actions allow determining the masses at dynamic model nodes.

6.11.2 *Preliminary Design*

Span and bay of the structure have been established by architectural requirements and validated by the structural engineer (Fig. 6.27).

The structural system naturally results from the pretty simple architectural layout: transverse identical frames with two spans and longitudinal central and marginal frames. Infrastructure is provided with perimeter walls, internal structural walls and some internal columns (Fig. 6.28).

The goal of preliminary design is to determine the proportions of structural components based on simple empirical formula and on constructive rules through a trial-and-error process. Initial sizes of structural elements will be validated by checking the fulfillment of "good seismic configuration" requirements.

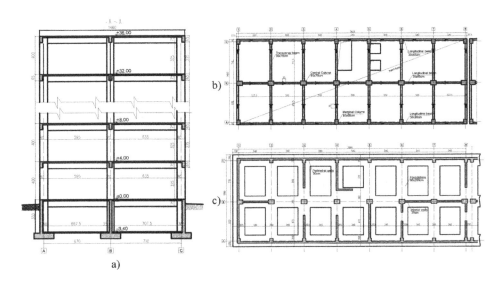

Figure 6.28 a) Building cross section. Structural layout of b) current floor and c) basement

Table 6.1 Preliminary cross section of columns

Floor	Marginal Column	Central Column
1	80 × 55 (cm × cm)	95 × 70 (cm × cm)
2	80 × 55 (cm × cm)	95 × 70 (cm × cm)
3	80 × 55 (cm × cm)	95 × 70 (cm × cm)
4	65 × 45 (cm × cm)	95 × 70 (cm × cm)
5	65 × 45 (cm × cm)	75 × 50 (cm × cm)
6	65 × 45 (cm × cm)	75 × 50 (cm × cm)
7	65 × 45 (cm × cm)	75 × 50 (cm × cm)
8	65 × 45 (cm × cm)	75 × 50 (cm × cm)
9	65 × 45 (cm × cm)	75 × 50 (cm × cm)

First Attempt

- *Slab thickness*: 14 cm (phonic insulation requirement)
- *Beams*: Height: $h_b \approx L/10$; Width: $b_b \approx h_b/2$ (for a good shear behavior)
- Chosen: – longitudinal beams 30 × 70 (cm × cm)
            – transverse beams 25 × 55 (cm × cm)
- *Columns* – staggered columns' shape was initially chosen, according to the variation of axial force as follows (Table 6.1)
- *Structural analysis*

According to the equivalent seismic forces approach, an elastic model was generated. For modal elastic static analysis nominal sectional stiffness as for whole (gross) concrete section $(E_c I_c)$ was accepted while for determining the lateral displacements the nominal section stiffness was reduced by a factor 0.50 for taking into account (approximately) the sections' cracking. Current software for static and dynamic analysis was used for

Table 6.2 Inter-story drifts for the first preliminary designed structure

Story	Elastic displacements		SLS drifts			ULS drifts		
	x	y	x	y	Limit	x	y	Limit
1	0.008	0.007	0.0050	0.0044	0.008	0.0031	0.0088	0.025
2	0.021	0.017	0.0093	0.0071	0.008	0.0163	0.0125	0.025
3	0.036	0.029	0.0107	0.0086	0.008	0.0188	0.0150	0.025
4	0.052	0.040	0.0114	0.0079	0.008	0.0200	0.0138	0.025
5	0.067	0.052	0.0107	0.0086	0.008	0.0188	0.0150	0.025
6	0.080	0.063	0.0093	0.0079	0.008	0.0163	0.0138	0.025
7	0.090	0.071	0.0071	0.0057	0.008	0.0125	0.0100	0.025
8	0.097	0.078	0.0050	0.0050	0.008	0.0088	0.0088	0.025
9	0.100	0.081	0.0021	0.0021	0.008	0.0038	0.0038	0.025

determining the modal parameters – vibration periods (three modes) and eigenvectors – and the inter-story drifts.

The modal analysis showed that the first two vibration modes are translations and the third is torsion. The vibration periods are: T1 = 1.48 sec, T2 = 1.27 sec and T3 (torsional) = 1.14 sec. So, from this point of view, the chosen structure met the recommended conditions.

Next step was to check the fulfillment of drift requirements for Service Limit State (SLS) and for Ultimate Limit State (ULS). The inter-story drifts calculated by the program are shown in Table 6.2.

From Table 6.2 results that the structure doesn't fulfill the drift requirements at SLS. Consequently, new sizes of structural elements will be checked.

Second Attempt

The newly chosen sizes of the structure's members are:

- Transverse beams: 30×75 (cm $\times$ cm)
- Longitudinal beams: 30×65 (cm $\times$ cm)
- Central columns: 70×95 (cm $\times$ cm) (constant over the whole building height)
- Marginal columns: 60×80 (cm $\times$ cm) (constant)

With the new seizes of the structural elements the following dynamic characteristics were determined:

- First mode $T_1 = 1.18$ sec (translation)
- Second mode $T_2 = 1.14$ sec (translation)
- Third mode $T_3 = 1.03$ sec (torsion)

The inter-story drifts of the new structure are shown in Table 6.3 and compared with admissible values.

From the Table 6.3 follows that the new structure's lateral stiffness complies with the drift control requirement.

The new structure can now be properly analyzed, designed and detailed.

Table 6.3 Inter-story drifts of the new structure

Floor	Elastic displacements		SLS drifts			ULS drifts		
	x	y	x	y	Limit	x	y	Limit
I	0.006	0.007	0.0038	0.0044	0.008	0.0023	0.0088	0.025
2	0.015	0.017	0.0064	0.0071	0.008	0.0113	0.0125	0.025
3	0.024	0.028	0.0064	0.0079	0.008	0.0113	0.0138	0.025
4	0.034	0.038	0.0071	0.0071	0.008	0.0125	0.0125	0.025
5	0.042	0.047	0.0057	0.0064	0.008	0.0100	0.0113	0.025
6	0.048	0.055	0.0043	0.0057	0.008	0.0075	0.0100	0.025
7	0.054	0.062	0.0043	0.0050	0.008	0.0075	0.0087	0.025
8	0.058	0.066	0.0029	0.0029	0.008	0.0050	0.0050	0.025
9	0.060	0.070	0.0014	0.0029	0.008	0.0025	0.0050	0.025

6.11.3 Structural Analysis and Design

Analysis is performed for the new structure loaded with vertical (gravity) loads and horizontal seismic equivalent forces, grouped in different combination according to the code. For "accurate" analysis more exact sectional stiffness is accepted namely: $(0.8E_cI_c)$ for columns and $(0.6E_cI_c)$ for beams where I_c is the inertial moment of concrete gross section.

The slab-beam system is analyzed using common elastic finite elements software. Different load patterns for live load which lead to maximum/minimum moments in slab panels are considered and, then, bending moment envelope has been determined. In a second step, some moment redistributions were accepted within a limit of 30% of elastic moments. Reinforcement amounts are within the economic domain, reinforcement percentage being between 0.7% and 1.4%.

Beams design has been performed according to the capacity method principles. The longitudinal reinforcement area was determined with maximum/minimum bending moments in potential plastic sections and, then, the beams were accordingly detailed. With the effective longitudinal reinforcement area flexural capacity of critical sections with over-strength and the associated shear force V_{Ed} have been determined. Transverse reinforcement (stirrups area and spacing) was provided to resist V_{Ed}. Stirrup spacing was checked for compliance with requirements related to preventing buckling of longitudinal rebars. Beams design and detailing are synthesized in Figure 6.29 and Tables 6.4 and 6.5.

According to the Capacity Method, columns are designed to respond elastically to the seismic action. For this purpose, bending moments in columns generated by the equivalent seismic forces are magnified with beam over-strength factor (see sub-chapter 6.5) and with a supplementary factor equal to 1.3 (Table 6.4 and 6.5). To the axial force determined from the load combination which involves equivalent seismic force and gravity loads is added (or reduced) the shear force associated with plasticization of beams' end section (Table 6.6 and 6.7).

Columns' reinforcement area was determined by using appropriate software. For central columns following reinforcement percentage are obtained: 1.8% at the base and 1.32% for sections in elevation. Central column's reinforcement percentages are 1.62% at the base and 1.2% for the rest (Fig. 6.30).

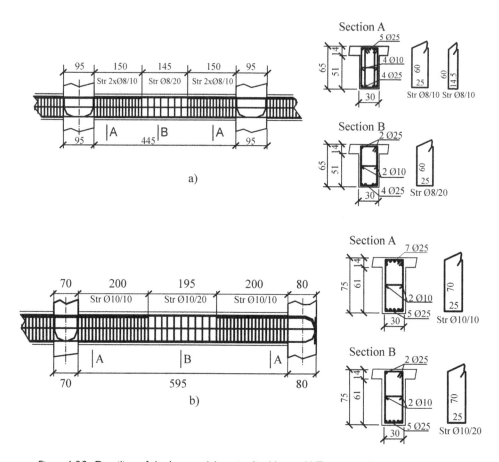

Figure 6.29 Detailing of the beams a) Longitudinal beam, b) Transverse beam

Table 6.4 Design elastic moments and associated shear force for longitudinal frame

Floor	Moments M_{ed} [kNm] Negative	Positive	Longitudinal reinforcement ratio [%] Section upper side	Section lower side	Design shear force V_{as}	Transverse reinforcement ratio (potential plastic zone)
1	389	350	1.26	1.01	198.45	0.67%
2	460	380	1.51	1.01	220.50	0.67%
3	462	360	1.51	1.01	220.50	0.67%
4	432	345	1.26	1.01	198.45	0.67%
5	384	300	1.26	1.01	198.45	0.67%
6	352	265	1.01	0.75	154.35	0.52%
7	260	160	0.75	0.75	132.30	0.52%
8	190	100	0.75	0.75	132.30	0.52%
9	142	68	0.75	0.75	132.30	0.52%

Table 6.5 Design elastic moments and associated shear force for transverse frame

| Floor | Moments M_{ed} [kNm] | | Reinforcemet ratio [%] | | Design shear force V_{as} | Transverse reinforcement ratio (potential plastic zone) |
	Negative	Positive	Section upper side	Section lower side		
1	578	300	1.40	0.93	157.50	0.52
2	675	450	1.63	1.17	189.00	0.52
3	676	450	1.63	1.17	189.00	0.52
4	635	430	1.63	1.17	189.00	0.52
5	570	430	1.40	1.17	173.25	0.52
6	490	330	1.40	0.93	157.50	0.52
7	4000	180	0.93	0.70	110.25	0.33
8	307	150	0.93	0.70	110.25	0.33
9	238	154	0.70	0.70	94.50	0.33

Table 6.6 Design moments in central column

| Floor | Elastic Moments [kNm] | | Amplification Factor | | Design Moments M_{ed} [kNm] | |
	Principal Direction	Secondary Direction	Principal Direction	Secondary Direction	Principal Direction	Secondary Direction
1	1853	1153	1.07	1.03	2587.53	1543.87
2	861	791	1.05	1.01	1175.27	1038.58
3	600	651	1.07	0.02	836.93	16.93
4	449	556	1.02	1.02	596.32	737.26
5	424	460	1.16	1.04	639.68	621.92
6	417	380	1.00	1.08	542.45	531.35
7	390	324	1.26	0.15	638.82	62.21
8	338	257	1.82	1.35	801.83	451.36
9	201	200	2.52	1.35	658.48	351.00

Table 6.7 Design moments in marginal column

| Floor | Elastic Moments' [kNm] | | Amplification Factors | | Design Moments M_{ed} [kNm] | |
	Principal Direction	Secondary Direction	Principal Direction	Secondary Direction	Principal Direction	Secondary Direction
1	1035	652	1.07	1.03	1445.27	873.03
2	608	540	1.05	1.01	829.92	709.02
3	454	477	1.07	0.02	633.28	12.40
4	373	424	1.02	1.02	495.38	562.22
5	310	363	1.16	1.04	467.69	490.78
6	290	294	1.00	1.08	377.24	411.10
7	270	218	1.26	0.15	442.26	41.86
8	236	139	1.82	1.35	559.86	244.12
9	235	124	2.52	1.35	769.86	217.62

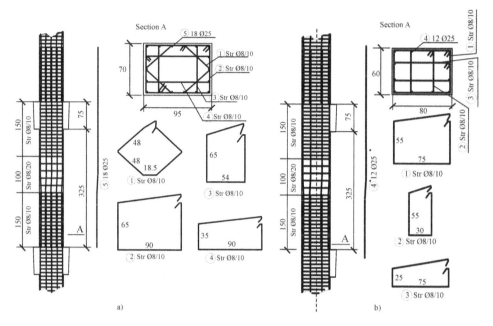

Figure 6.30 Columns' detailing a) Central column, b) Marginal column

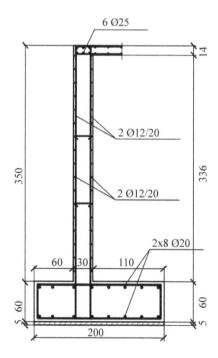

Figure 6.31 Peripheral wall reinforcement

Infrastructure Analysis and Design

Consisting of a system of walls, the infrastructure located at the basement level shows a substantial lateral stiffness in comparison with the superstructure.

For analysis, the infrastructure components were modeled as a system of elastic finite elements. The soil was modeled with Winkler's springs. Loads acting on the infrastructure are: the axial forces in columns due to gravity loads, the flexural capacities of columns' base, shear and axial forces associated to the dissipating mechanism. A common FEM software has been used to calculate the internal forces in infrastructure components and, then, to determine the reinforcement area – horizontal and vertical. The resulting reinforcement percentages range between 0.2 and 0.4% (Fig. 6.31).

Conclusions

The chapter deals with theoretical and practical aspects raised by the modern seismic analysis, design and detailing of reinforced concrete frame structures. It is shown how the concept of structural system could be applied in the case of framed concrete buildings. A case study is included. It presents in detail the structural design steps and detailing drawings of a nine-story building located in a high intensity seismic zone.

Chapter 7

Structural Wall Systems

Abstract

Systems with structural walls show important advantages from the point of view *resistance* to lateral forces, *lateral stiffness*, and construction (execution) as compared with frames. Behavior of wall systems and of their most used components – cantilever and coupled walls – is examined. According to the overall cantilever behavior of the wall system under seismic actions, rules for *proper conformation* are formulated and recommendations for their conceptual design are defined. Due to their layout and proportions the structural walls systems require specific analysis approach. Simplified and accurate analysis of structural walls is presented. Design and detailing of structural wall systems including superstructure and infrastructure with their components and mutual interaction are analyzed. Case study about a 19-story residential building with two-level basement illustrates the design and detailing steps of a typical structural walls system.

7.1 General

Multi-story frame structures subjected to seismic actions have a limited use for tall buildings due to already mentioned specific disadvantages. These disadvantages are related to the following features:

- Since the frames are made by relatively flexible elements (beams and columns) drift control requirement is difficult to fulfill for high-rise buildings, located in high intensity seismic zones.
- For high-rise buildings, the columns' and beams' sizes at the bottom zone of the building are large. Consequently, an important part of the building area at its bottom floors is not functional.

Searching for solutions to extend the use of reinforced concrete for high- and super-high-rise buildings the idea, which arose in the early 1950s, was to convert partitions (non-structural components) into *structural walls*. The walls have a high "natural" stiffness due to their size and can be easily provided with substantial resistance capacity (Park, R. & Paulay, T. 1975).

Thus, reinforced concrete structural walls have a double function:

- They are structural (vertical) elements aimed to support and transfer vertical and horizontal loads;
- Delimit the building rooms.

Systems with structural walls show important advantages from the point of view of *resistance* to lateral forces, *lateral stiffness*, and *construction* (execution) as compared with frames. They allow the use of flat slabs for floor structure; formworks and detailing are substantially simplified with evident labor and overall cost savings. The architectural layout is free of massive elements (columns and beams).

On the other hand, obvious structural requirements oblige to keep the walls continuous over the building height. This is a significant limit of using the structural walls for multistory buildings since the architectural layout has to be kept the same for all building stories during its lifetime (minor changes can be made only within cells delimited by structural walls).

For this reason, the structural wall systems are suitable for apartment buildings, hotels and other similar buildings which keep their function and layout unchanged during lifetime.

7.2   Types of Structural Walls

Cantilever walls. An ideal solution for structural wall systems is to provide them with compact, solid units which are the *cantilever walls* (Fig. 7.1). They can be easily executed with planar moulds and their detailing is simple. The load paths of cantilever walls are direct and unambiguous.

The most frequent cross section of cantilever walls is rectangular with unusual proportions: height (equal to wall width) of several meters and width (wall thickness) of about 20 to 50 cm. In order to prevent buckling of the compressed zone, to ensure resistance to compressive stresses and to ease the detailing, thin walls can be provided with reinforced boundaries. Intersected walls provided on two perpendicular directions lead to more complicated shapes of the cross section.

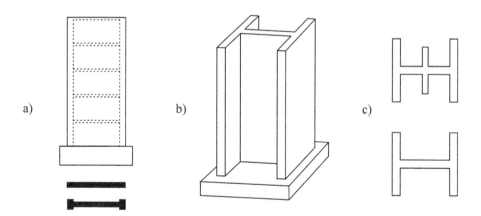

Figure 7.1 Cantilever structural walls

Structural walls with openings. Need to include in walls windows and doors frequently oblige to provide them with openings.

Walls with one or more vertical rows of openings are called *coupled walls* They can be regarded as two or more cantilever walls coupled by beams (Fig. 7.2a). Solutions with staggered openings have been also implemented (Fig. 7.2b). They have an excellent stiffness but, sometimes, the internal force paths are less evident and require careful analysis (Fig. 7.2c).

Special types of structural walls Both cantilever and coupled walls can be performed by *cast-in-place* or by *prefabricated* concrete. Rectangular prefabricated units called *large panels* are interconnected to realize cantilever or coupled structural walls (Fig. 7.3a).

Braced reinforced concrete frames behave similarly to structural walls (Fig. 7.3b).

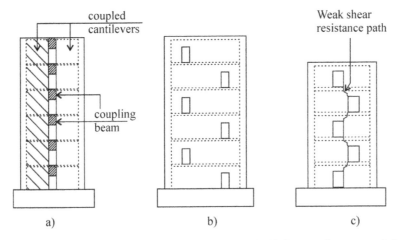

Figure 7.2 Structural walls with openings a) Coupled walls, b) Staggered openings, c) Staggered openings with weak shear resistance path

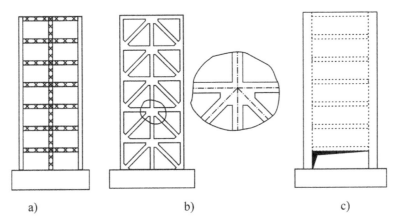

Figure 7.3 Special types of structural walls: a) Prefabricated large panels wall, b) Bracing frame behaving similar to walls, c) Wall with soft story

Structural walls supported by columns on the first floor have been used in the past, in order to obtain architectural freedom (large rooms) on this level. They are walls with soft and weak story (Fig. 7.3c). This type of structural wall evidenced very poor seismic behavior and, normally, should be avoided.

7.3 Behavior of Wall Systems

7.3.1 General

Wall systems are complex structures involving 2D components – cantilever and coupled walls – with substantial difference in their elastic and post-elastic seismic response.

Having width of several meters and thickness ranging from about 20 to about 50 cm, cantilever walls show specific behavior peculiarities depending upon their aspect ratio and upon their cross section layout and size.

Currently used in wall systems are also the coupled walls with one or more uniform row(s) of openings. Coupled walls are themselves complex units involving two different types of components – the solid walls and the coupling beams – each of them with its specific behavior peculiarity. The openings' size as compared with story height determine the aspect ratio of the coupling beams and, accordingly, their stiffness and specific behavior. Currently, they are short and deep beams showing a shear dominated response. The coupled wall overall behavior, somehow similar to that of frames, is strongly influenced by the coupling beam/wall stiffness ratio ("coupling degree") and by the detailing of these two components.

Wall system elastic and post-elastic seismic response peculiarities can be deduced from the behavior of its components: cantilever and coupled walls.

Seismic behavior of its structural components and of the wall system itself will be briefly examined below.

7.3.2 Behavior of Cantilever Wall

Elastic and post-elastic behavior of solid cantilever walls is strongly influenced by their aspect ratio H/l which synthetically expresses the contribution of bending moments as compared with that of shear force in total element deformation. From that point of view, two limits exist (Fig. 7.4):

- Slender walls – flexural dominated
- Squat walls – shear dominated.

Slender walls are specific to medium- and high-rise buildings. Their response to seismic horizontal forces is that of a cantilever which is a statically determined element. Maximum internal forces, bending moment and shear force, have their maximum magnitude toward the wall base. Accordingly, plastic deformations will be expected at this zone along a *potential plastic length* l_p (see Chapter 2, paragraph 2.4.3). For structural walls l_p length is given by codes. Normally, it is accepted that the potential plastic zone is spread over the entire one or two story height (Fig. 7.5).

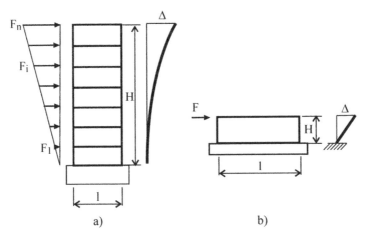

Figure 7.4 Slender and squat walls a) Slender wall, b) Squat wall

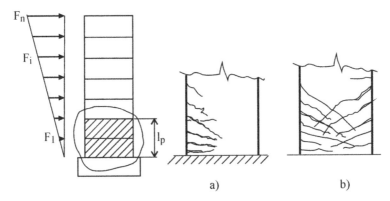

Figure 7.5 Potential plastic length a) Cracking under monotonic loading, b) Cracking under reversal loading

7.3.3 *Coupled Wall Behavior*

Coupled walls involve two components which show very different behavior peculiarities: walls and coupling beams. The specific global seismic response of coupled wall is a result of its components' behavior. Normally, the coupling beams are short and deep elements so that their behavior is shear dominated. However, global seismic behavior of coupled walls is similar to that of frames. So, plastic deformations occur initially in beams, spreading over the wall height from bottom to the top. The failure mechanism is completed through plastic deformations at walls' bottom.

Behavior of Coupling Beam

The symmetrical coupled walls respond to horizontal forces by deforming as shown in Figure 7.6a. Accordingly, the coupling beam deformed shape is as shown in Figure 7.6b.

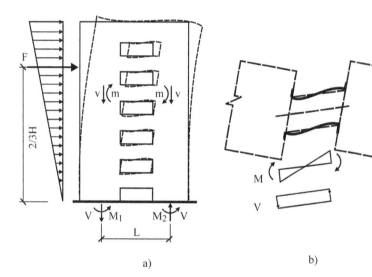

Figure 7.6 a) Symmetrical coupled walls deformation b) Deformation of coupling beams

This means that the coupling beams of symmetrical wall systems respond to seismic action by an anti-symmetrical deformation.

In order to simulate the behavior of coupling beams, according to their specific deformation shape, Thomas Paulay conceived in the late 1970, test units and testing procedure as shown in Figure 7.7.

Extensive experimental researches have been carried out on such specimens, which yielded essential information on their behavior. Further researches completed the picture of the behavior of these elements, loaded up to failure, by monotonic as well as hysteretic loading (Paulay, T. 1969).

A. Behavior under monotonic loading

A coupling beam, having an aspect ratio $l/h = 1 \ldots 3$, loaded monotonically up to failure shows the following *behavior stages*:

- *Elastic un-cracked* – for shear forces of small intensity the beam is un-cracked; internal forces distribution corresponds to that of an elastic element.
- *Elastic cracked* – the first cracks are of flexural type; they are, generally, located in the sections "weakened" by the stirrups. Initialized by these cracks, inclined cracks are developed. The inclined cracks significantly change the stress distribution within the element: both longitudinal reinforcements (top and bottom) become tensioned along the whole element length. Accordingly, the coupling beam behaves similarly to two interconnected short cantilevers (Fig. 7.8).
- *Post-elastic (cracked)* – at a certain magnitude of the external incremental loading, non-linear phenomena occur throughout the beam i.e. plastic strains of compressed concrete struts and yielding of longitudinal and/or transverse reinforcements. The

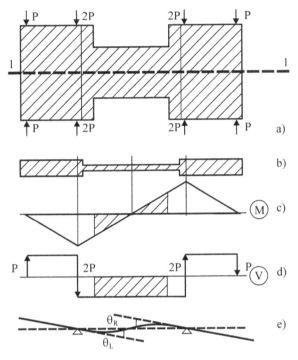

Figure 7.7 Testing specimen for investigating coupling beam behavior a) Front view, b) Top view, c) Bending moment diagram, d) Shear force diagram, e) Deformations

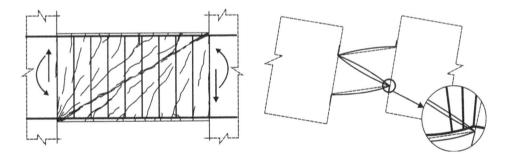

Figure 7.8 Behavior of cracked coupling beams

path of plastic deformations is strongly influenced by the ratio between flexural and shear strength of the beam. Other parameters could have also their substantial contribution.

- *Ultimate* – the beam fails due to one or more combined effects: crushing of concrete of inclined struts, failure of stirrups along a diagonal critical crack, loss of anchorage of longitudinal bars.

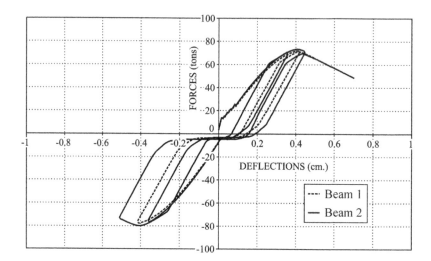

Figure 7.9 Force-displacement relationship with "pinching effect" in coupling beams

B. Hysteretic behavior

The behavior to high-intensity cyclic reversed loading ("hysteretic behavior") depends upon the loading history and upon the ratio between the maximal displacement and the ultimate one (the "degree of penetration" into post-elastic range). The main phenomena, associated to this behavior, are the following:

- Gradual deterioration of concrete struts due to repeated cracking (each concrete fiber is successively loaded in compression and tension) and progressive deterioration of concrete covering.
- Slip of longitudinal bars due to gradual loss of bond.
- Potential buckling of longitudinal bars especially when the stirrups don't stabilize them correctly
- Mutual slip of beam portions separated by cracks that are not completely closed from one loading cycle to another.

At the element scale these phenomena lead to a progressive diminution of shear strength and/or stiffness as well as the energy absorbing capacity. The stiffness degradation is substantial just after reversing the load and is visualized as the *pinching effect* of the force/displacement curve (Fig. 7.9).

Global Behavior of Coupled Wall

A coupled wall is a statically indeterminate structure that requires itself a specific analysis, much more complicated than the cantilever walls.

The beams interconnect the vertical walls ensuring a "coupling effect" that depends upon the ratio of beam and wall stiffness.

If the walls are interconnected at each level only by the slab, which can be assumed to be infinitely flexible, they act as quasi-independent cantilevers (the slab, acting as a

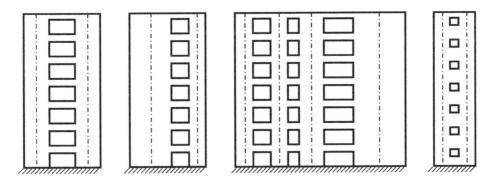

Figure 7.10 Walls with different opening sizes

diaphragm, coordinates the walls' lateral displacements). Large openings (for doors, for example) lead to weak and flexible coupling beams. Thus, the coupling effect of such beams is negligible and the wall with large openings can be considered as a set of cantilever walls.

On other limit, a wall with small openings (for small size windows, for example) behaves like a unique wall even though with a certain loss of stiffness.

Between these two limits, lie most walls with openings for which the coupling effect of the beams is more or less important, depending upon the size of openings that determine the coupling beam aspect ratio and, accordingly, its stiffness and resistance (Fig. 7.10).

Let us consider two walls interconnected with coupling beams. Their mutual slip is partially impeded by the coupling beams. Thus, the coupling beams are subjected to shear forces, proportional to the shear of the assembly of walls. Beam shear force is transferred to each wall where it becomes axial force – compression in one wall and tension in another.

In other words, the *overturning moment* at each level of the coupled wall is balanced by the bending moments in each wall as well as by the couple of axial forces (compression and tension) developed in walls. In their turn, the axial forces in walls are the sum of shear forces of coupling beams above the considered level (Fig. 7.11):

$$M_0 = M_1 + M_2 + N \cdot d \tag{7.1}$$

$$N = \Sigma V_{\text{beam}} \tag{7.2}$$

Two important specific behavior features result from these statements:

1. *For the coupling beams.* The walls are bearing elements for the coupling beams. Because of axial forces, one end of the beam is lifted (due to wall elongation generated by tensile axial force) while the other is displaced down (by the compressed wall). Because the coupling beams are very rigid, the vertical displacements of their ends strongly influence the magnitude of internal forces.

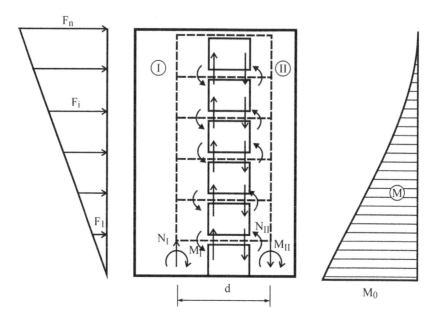

Figure 7.11 Internal forces in coupled walls

2. *For the coupled wall.* The magnitude of internal forces in coupling beams is proportional to the shear force of the coupled wall. Since the wall shear force increases from top toward the bottom, same variation of the beams' internal forces occur. At a certain loading level, plastic deformation will be developed within the coupling beams (generally weaker than the walls) starting from the base of the structure and progressing toward the top. When the lateral loads increase, some of the beams, at the structure bottom, exceed their plastic deformation capacity and progressively fail. The coupled wall behavior (elastic, post-elastic and progressive failure) is illustrated by Figure 7.12.

It can be noticed that, through progressive yielding of the coupling beams, a considerable amount of energy is dissipated. Thus, the overall ductility of the coupled walls is substantially enhanced in comparison to the cantilever walls.

7.3.4 Wall System Behavior

Wall systems are space (3D) structures made of vertical cantilever and coupled walls rigidly inter-connected, on each floor, by horizontal diaphragms.

Under lateral forces the whole system acts as a cantilever. Its spatial seismic response involves three components: two translations along principal axes of the system and a global torsional rotation (torque).

For low magnitude of lateral forces the wall system responds elastically. The system global deformation depends upon its elastic stiffness – translational and torsional. The system stiffness is a function of the magnitude and of in-plan distribution of

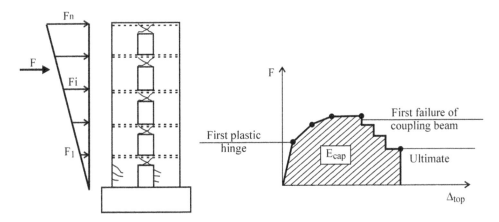

Figure 7.12 Coupled wall behavior up to the failure

its components' rigidities (cantilevers and coupled walls). Synthetically one can say that the translational stiffness along a principal axis of the system is the sum of elastic stiffness of its components along the considered direction. The global torsional rotation occurs about a vertical axe passing through the *stiffness* (or *rigidity*) *center* of the system which is the centroid of elastic stiffness of the system components (see sub-chapter 7.6). The system torsional elastic stiffness is the sum of polar moments of its components' stiffness about the stiffness center. These somehow complicated statements will be thoroughly examined and explained within subchapter 7.6.

The system responds elastically until the sectional resistance is reached at the element with lower capacity/demand ratio. This could be either a cantilever or a coupling beam. Cantilever wall with plastic zone at its bottom is no longer active in carrying force increment but keeps constant its maximum lateral force (ideal elastic-plastic behavior is assumed). Plasticization of the first element creates unbalanced distribution of rotational tendency of the system. Consequently, the rotational center (stiffness centroid) migrates toward a new position. When the plasticization of all structural walls (cantilever and coupled structural walls) is reached the system rotates about a vertical axis passing through the centroid of *lateral shear resistance* of its components. We call this point the *resistance center*. Thus, under incremental lateral forces the structural system with walls responds through translations about two principal axes and rotates about vertical axes situated between the stiffness center and the resistance center.

The elastic-plastic response above described is effective provided that all components of the system are ductile, able to develop substantial post-elastic deformations without failure.

7.4 Conceptual Design of Structural Wall Systems

When subjected to lateral forces, the whole building acts as a vertical cantilever embedded into soil through the *foundation system* or *infrastructure*.

According to the overall cantilever behavior of wall system under seismic actions, rules for their *proper conformation* can be defined. These rules allow ensuring a *favorable response* to seismic actions even from preliminary design stages.

Seismic favorable response is characterized by a stable and controlled behavior with limited unfavorable phenomena (excessive lateral displacements and damage of structural and nonstructural components, uncontrolled $P-\Delta$ effect, etc.) and by predictable post-elastic response.

Within conceptual design, the wall distribution in plane and elevation, the conformation of the infrastructure, choice of solution for slab system, etc. are keys for obtaining structural systems that ensure an optimal compromise between functional, economic and execution requirements, obtaining also a favorable structural response.

Rules for obtaining these requirements can be defined as follows:

(i) Symmetrical wall distribution

Within building, it is advantageous to provide structural walls symmetrically regarding the two principal axes. The symmetry concerns both geometrical shape and stiffness of the walls. Structures with unsymmetrical walls distribution have the tendency to rotate around the torsional center generating a global *torsional moment (torque)*. The general torsion of the structural system overloads the walls especially those located at the building periphery. The symmetrical distribution of walls reduces or eliminates this effect.

(ii) Comparable stiffness of wall system according to both principal directions

During severe seismic actions, plastic deformations are developed in certain walls that, sometimes, can fail. Even for symmetrical structures the yielding of similar walls is not simultaneous but sequential depending upon the ratio between their *real* resistance and *effective* load. Accordingly, global torsion occurs even in perfect symmetrical structures. This tendency must be balanced by moments generated within "sound" walls still existing on both directions.

Consequently, structures with structural walls (rigid system) on one direction and frames (flexible system) on the other are not recommended. During severe earthquakes the walls situated at greatest distance from the torsional center are the most likely to fail. Consequently, the torsional stiffness of the system dramatically drops, increasing the tendency of unbalanced global twist. Since the frames are flexible, they don't participate significantly in carrying the torsional moment and the remaining walls are drastically overloaded. So, the structure has no strength reserves and it is possible that failure of one wall leads to progressive failure of the whole system (unstable process).

(iii) Providing walls on two directions so that significant residual torsional stiffness of the system is assured even when some walls fail

The existence of walls with similar stiffness on both principal directions is not sufficient for assuring a stable behavior regarding the global torsion.

As an example, the structure of Figure 7.13 has similar stiffness in both principal directions. A single wall, provided along the building longitudinal centerline, doesn't

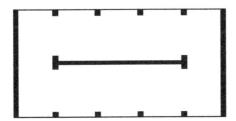

Figure 7.13 Structure with single longitudinal wall

participate in the torsional stiffness of the system. For this reason the failure of one transversal wall makes the system unstable regarding the torsional resistance (and torsional stiffness).

(iv) Providing horizontal diaphragms with high stiffness and enough strength capacity

The walls situated in one or more direction(s) within a building become a structural system only if their displacements are *coordinated* by stiff and resistant horizontal elements that behave like *diaphragms.*

Generally, the slabs are designed to carry and transfer vertical loads. Because of their proportions, the slabs have substantial natural resistance and stiffness when loaded in their plan. Accordingly, in almost all cases the slab acts like a diaphragm for the wall system without any special measures.

The diaphragms are loaded in their plan by seismic forces (inertial forces applied to masses dynamically moved by seismic action) and are supported by the structural walls. Acting as horizontal beams loaded in their horizontal plan, the diaphragms are subjected to high magnitude bending moments and shear forces.

When the wall spacing is moderate and the slabs are without large openings the diaphragm function is fulfilled without special measures.

Special measures for ensuring diaphragm effect are necessary when:

(a)   Distance between walls is significant
(b)   The slab has large openings
(c)   The slab is prefabricated (for example pre-cast hollowed slabs)

(v) Choice of system configuration that leads to favorable loading of foundations

The structural wall systems transfer to the infrastructure and foundations lateral loads in a different way than those transmitted by the frame systems in terms of intensity and distribution.

The frame structures present a uniform distribution of columns regarding the position and strength capacity. According to the *capacity design method* the infrastructure is loaded by the capacity of the superstructures. This means that the frame system infrastructure is loaded uniformly by the superstructure (vertical and horizontal

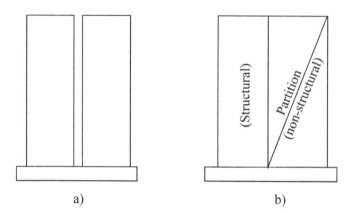

a)   b)

Figure 7.14 Solutions to adapt stiffness of large width walls a) Split of a large width wall, b) Wall
having only partially structural function

forces and moments with close values to each other and short distance between
them). Consequently, the frame system infrastructure is favorably loaded since no
high magnitude loads have to be transferred over great distance.

In contrast with frame structures, the wall systems generally consist of more limited
number of components (i.e. structural walls) having variable strength and stiffness from
one to another. Accordingly, the infrastructure has to transfer to the soil substantial
lateral loads non-uniformly distributed.

From very beginning of the design process, the structural engineer has to seek the
best compromise between functional requirements (which imposes the wall position)
and advantageous loading of each wall and of the infrastructure. As a general rule,
the structural walls system has to load the infrastructure as uniformly as possible, and
the axial force of walls shall be as great as possible in order to stabilize the foundation
system.

(vi) Avoid great differences between wall strength capacities

The architectural constraints often lead to substantial difference between walls' dimensions. For example, façade walls or the walls that separate apartments could have
large width, while the walls situated within apartments are, normally, of limited size.
Consequently, the infrastructure has to transfer to the soil very non-uniform loads
resulting from walls' strength capacity. A possible solution for obtaining walls with
close strength capacities is to split large width walls through joints. Another option
is to provide structural walls with limited width even when the architectural layout
would allow much wider walls (Fig. 7.14).

(vii) Avoid wall section with substantial non-symmetry of resistance

Walls that have T-shape cross-section with large flanges have very different capacities
for reversed loadings. When the flanges are tensioned a large amount of reinforcing
bars is mobilized, which leads to great wall resistance. (Actually, this resistance is

difficult to accurately quantify since the active flange width within tensioned zone depends upon many parameters). For reversed moment, the wall capacity is much lower. Accordingly, the infrastructure is subjected to loads having a magnitude that depends upon the sign of horizontal forces. It can be overloaded in some moments of the seismic action.

(viii) Ensure appropriate magnitude of axial force in walls

The floor slabs transfer their vertical loads to the walls. Slab vertical reaction becomes axial force in walls. Accordingly, the slabs' configuration strongly influences the intensity of axial forces in walls.

When conceiving the slab system pattern (slab spans, distribution of supporting elements), the designer has to take into account that the selected solution has to transfer reactions on the walls so that their axial force is:

• Of high enough intensity for ensuring reasonable vertical reinforcement amount of the wall
• Of an intensity that ensures wall ductile behavior $(n = N/Afc \leq 0.2\text{–}0.3)$.
• Advantageous for infrastructure (axial force has a stabilizing effect).

7.5 Analysis of Wall Systems

7.5.1 Elastic analysis of Wall Systems

Wall systems are complex spatial structures involving planar (2D) components. For quantifying their full elastic-plastic seismic response, nonlinear finite element software would be required. Moreover, reliable constitutive laws for cantilever 2D elements, for coupling beams having different aspect ratio and for different types of coupled walls are scarce and cover only limited domain. For these reasons, "exact" post-elastic seismic analysis of wall systems is a tremendous task nowadays performed for research purposes only for particular cases.

Currently, seismic analysis of structural wall systems uses the equivalent seismic force approach being performed with software packages with elastic finite elements. Most finite element programs allow integrating the stresses within element cross section and to obtain internal forces: bending moments, shear and axial forces.

Components of wall system – cantilever and coupled walls as well as the horizontal diaphragms – are modeled through shell type finite elements (Fig. 7.15).

For taking into account globally, in a conventional way, the post-elastic behavior of system components nominal magnitude of elasticity modulus, differentiated for specific elements (compressed cantilevers, tensioned cantilevers, squat walls, coupling beams, etc.) are recommended by codes and by literature. Another way to take into account, during design process, post-elastic behavior is by adjusting, when necessary and advantageous, the elastic moments by up to 30%, observing, in the same time, the overall equilibrium of the structure (see paragraph 3.2.5).

Proper post-elastic analysis of wall systems can be done on models of equivalent frame (see next paragraph). Cantilever walls are modeled in a simplified way as linear bars.

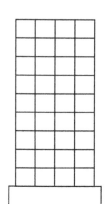
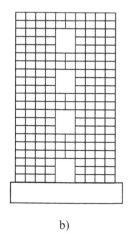

Figure 7.15 Finite element models for wall system components a) Cantilever wall, b) Coupled wall

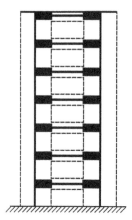
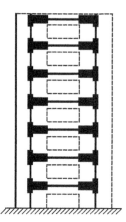

Figure 7.16 Equivalent frame of coupled walls

Method of equivalent frame. Although the coupled walls are structures with elements having proportions of deep beams or squat walls, they can be approximately modeled as *plane elastic frames*, by implementing some corrections that take into account the real behavior.

Geometric modeling accepts to concentrate sectional properties in centerline of walls and coupling beams. Within intersection of beams and walls, infinitely rigid elements are considered (Fig. 7.16).

Sectional properties are provided by codes, accepting as reference values those of un-cracked gross concrete area E_c, I_c, A_c. Corrections are then implemented for taking into account (approximately) the cracking of the concrete and other significant factors that influence the structural response. For example, the following equivalent sectional properties can be accepted:

- for compression (external) wall $(EI)_{eq} = 0.8(E_c I_c)$
  $(EA)_{eq} = (E_c A_c)$

- for tension (external) wall $(EI)_{eq} = 0.5(E_c I_c)$
  $(EA)_{eq} = 0.5(E_c A_c)$

- for internal walls $(EI)_{eq} = 0.7(E_c I_c)$
  $(EA)_{eq} = 0.7(E_c A_c)$.

Sectional stiffness of the coupling beam is strongly influenced by the deformations due to shear. Accordingly, the following expression can be accepted for its equivalent flexural stiffness:

$$(EI_b)_{eq} = 0.2 E_c I_c \left[1 + 1.5 \left(\frac{h_b}{l_b} \right)^2 \right] \leq 0.8 E_c I_c \tag{7.3}$$

The equivalent frame is loaded with lateral forces distributed similar to seismic or wind forces. Standard elastic analysis is then performed. Most computer programs for planar frames accept the assumption of "finite joint" (rigid zones where beams frame into the walls) and so, the geometric modeling of equivalent frame is easily taken into account. Due to the proportions specific to "deep beam", the coupling beam internal forces are influenced by the axial deformations of the walls (which are bearing elements for beams). Thus, proper analysis of equivalent frames requires computer programs based on a stiffness matrix procedure that automatically consider the axial deformations of the members.

For rapid verifications and for better understanding the analysis backgrounds of structural walls systems it is useful to know the pre-computer, traditional analysis approach. Codes and literature frequently refer to concepts and notions of this approach. Accordingly, the traditional analysis of wall systems will be presented within the next sub-chapters.

7.5.2 Post-elastic Analysis of Wall Systems

For practical purposes nowadays no post-elastic finite element software exists. When post-elastic analysis – pushover or time-history – is needed for special structures equivalent frame approach has to be used.

Design and detailing of walls structural systems is a trial-and-error process starting with modeling of structural components, then performing preliminary analysis, determining the design quantities according to capacity design method, proportioning and detailing of structure members. Having specific geometric and behavior features, each type of component of wall systems has to be thoroughly examined from the point of view of behavior, analysis, design and detailing. This will be presented within the next subchapters.

7.6   Simplified Analysis of Wall System

7.6.1   General

The wall systems are subjected to vertical (gravity) loads and to horizontal forces due to wind and seismic actions.

The analysis of structural walls system is made separately for two kinds of loads – vertical and horizontal.

Generally, gravity loads don't raise special problems. As already shown (see chapter 6), each slab panel transfers to its bearing elements (walls or beams) loads that act on *tributary areas* resulting from dividing the slab panel through 45° plans.

The total slab reaction on each wall, added along wall height, gives its axial force.

The simplified analysis examined within the present chapter refers to the wall system, considered as a whole, under lateral equivalent seismic loads. Thus, this is the conventional elastic analysis suitable for manual approach.

It is assumed that the system is provided with infinitely rigid horizontal diaphragms that coordinate the displacements of walls on each building story. Accordingly, on each story, the system has a rigid body displacement quantified through three independent components: linear displacements about two axes and a rotation (twist) about a vertical axis that passes through a point called *rigidity center* $CR(x_0, y_0)$. The rigidity centre is the point having the following attribute: if a horizontal force passes through this point, the system will have only linear displacements but no rotation.

For the *manual approach* further simplifications have to be accepted in order to reduce the number of unknown quantities. A substantial simplification results if we accept that the deformed *shape* of each structural component of the system (each wall) is the same. Consequently, instead of calculating the displacements at *each level*, it is enough to know the displacement at one *reference* level. System displacements at all other levels are proportional to those of the reference level. This means that, instead of 3 m unknown quantities (m is the number of building stories) only three independent quantities remain: the linear displacements and the rotation at reference level.

It is accepted that the reference level is either the top of the building or at a level situated at $0.8H$ (H is the total building height).

For the analysis the stiffness of each structural component (cantilever walls, coupled walls or frames) has to be defined in an appropriate way. Generally the stiffness is defined as the *generalized force* corresponding to a *unit displacement* on its direction. In our case, under lateral forces, each structural component acts as cantilever. The displacement is considered the deflection at the reference level (at the top or $0.8H$). Consequently, the general definition of the stiffness is adjusted in order to correspond to the accepted simplified approach. So, the stiffness of each element (cantilever wall, coupled wall or frame) is considered to be the *total lateral force* that produces a displacement equal to one at the reference level. The lateral force is distributed along the element height according to its nature: triangular, or constant (Fig. 7.17).

The simplified *system* analysis involves the following steps:

1.   Calculation of stiffness of each element in both directions ($k_{x,i}, k_{y,i}$)
2.   Compute the total lateral force acting on the system and its distribution over the building height
3.   Determine the coordinates of rigidity centroid

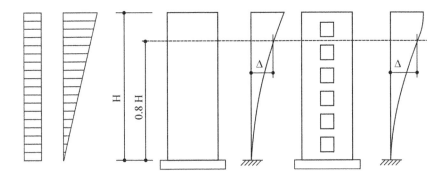

Figure 7.17 Definition of wall stiffness for simplified analysis

4. Determine the forces in each wall due to the *direct* effect of total lateral force, which induces lateral displacement
5. Determine the forces in each wall due to the *general twist*
6. Add, for each component, the two forces above determined
7. Perform the analysis, design and detailing of each component under lateral and gravity (axial) forces

7.6.2 Coordinates of Rigidity Centre

It is assumed that, for each component of the system $(1), (2), \ldots, (j)$, the stiffness about its principal axes is known k_{xi}, k_{yi} $(i = 1, 2, \ldots, j)$ (see below appropriate subchapter for each structural component).

Moreover, it is supposed that all components of the system have their cross-section principal axes *parallel* to each other. (This assumption will be further commented upon).

We impose on the system a displacement $\Delta y = 1$ (Fig. 7.18). Due to the horizontal diaphragm all elements of the system will have the same displacement. Accordingly, in each element a force will be generated which is, according to the definition, equal to the element stiffness. The result of these forces is the *system global stiffness* according to the axis y: $K_y = \sum_{i=1}^{j} k_{y,i}$.

Similarly: $K_x = \sum_{i=1}^{j} k_{x,i}$.

Since the imposed displacement is pure translation (without any rotation), the resultants K_x and K_y pass through the rigidity centre $CR(x_0, y_0)$. Writing that the moment of the result is equal to the moments of its components, about any point of the plane, the coordinates of CR are reached:

$$K_y \cdot x_0 = \sum_{i=1}^{j} k_{y,i} \cdot x_i \quad \Rightarrow \quad x_0 = \frac{\sum_{i=1}^{j} k_{y,i} \cdot x_i}{\sum_{i=1}^{j} k_{y,i}} \tag{7.4}$$

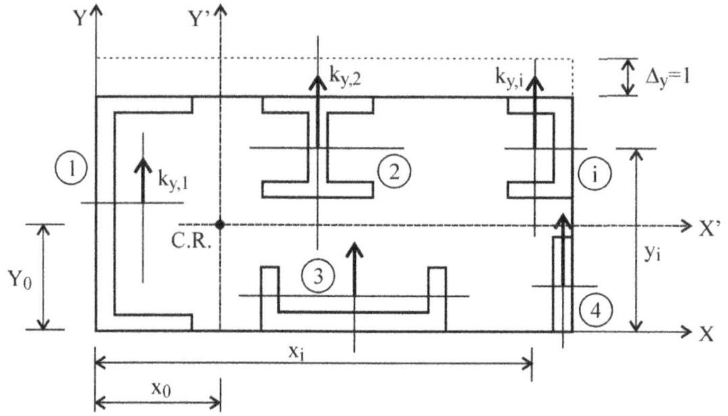

Figure 7.18 Coordinates of stiffness centre

and, similarly, $\Rightarrow y_0 = \dfrac{\sum\limits_{i=1}^{j} k_{x,i} \cdot y_i}{\sum\limits_{i=1}^{j} k_{x,i}}$ (7.5)

We suppose now that the total lateral force F acts on the system. The force F has two components F_x and F_y according to the principal axes of the system. (Taking into account the assumption of parallel principal axes of the components, the principal axes of the system are parallel to these axes too).

Reducing the external force F (or its components F_x and F_y) to the CR, the system is subjected to:

A.   The external force F (or its components F_x and F_y) which passes through the centroid CR and generates only linear displacements (or *translation*) ("*direct*" effect) and

B.   Global torsional moment (torque) $M_t = Fd = F_x d_x + F_y d_y$ (d, d_x and d_y are the distances of F, F_x and F_y to CR)

The two effects will be separately treated.

7.6.3   Direct Effect of the Lateral Force (Effect of Translation)

Let us suppose that the system is subjected to the external lateral load F_x that passes through the stiffness centroid CR. Consequently, the system will have a displacement Δ_x. Since elastic behavior of the system is accepted, the displacement Δ_x is proportional with the force F_x:

$$F_x = K_x \Delta_x \tag{7.6}$$

where K_x is the system stiffness.

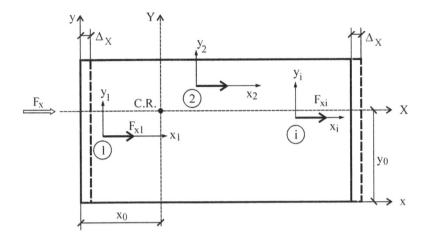

Figure 7.19 Direct effect of the lateral force

The same displacement Δ_x is imposed on all components of the system (assumption of infinitely rigid diaphragm). Thus, in each system component i, the force F_{xi} is developed proportional to its stiffness k_{xi}:

$$F_{xi} = \frac{k_{xi}}{K_x} F_x = \frac{k_{xi}}{\sum k_{xj}} F_x \tag{7.7}$$

Similarly, on y direction, results:

$$F_{yi} = \frac{k_{yi}}{K_y} F_y = \frac{k_{yi}}{\sum k_{yj}} F_y \tag{7.8}$$

Thus, *the lateral force in each direction is shared between structural components proportionally to their stiffness* (calculated for this direction).

7.6.4 *Effect of Rotation (General Twist)*

The external force F reduced to the centroid CR gives a force of same magnitude and direction and the global torsional moment $M_t = Fd$ (d is the distance between the force F and the centroid).

Due to the torsional moment, a general rotation θ occurs about CR so that each element is displaced according to a circle with the center in CR. It can be accepted that the displacement on the circle has about the same magnitude as that of tangent (Fig. 7.20):

$$\begin{cases} \Delta_i = \theta \cdot r_i \\[4pt] \Delta_{ix} = \theta \cdot Y_i \\[4pt] \Delta_{iy} = \theta \cdot X_i \end{cases} \Rightarrow \begin{cases} F_{ix}^{M_t} = k_{x,i} \cdot \Delta_{ix} = k_{x,i} \cdot \theta \cdot Y_i \\[4pt] F_{iy}^{M_t} = k_{y,i} \cdot \Delta_{iy} = k_{y,i} \cdot \theta \cdot X_i \end{cases} \tag{7.9}$$

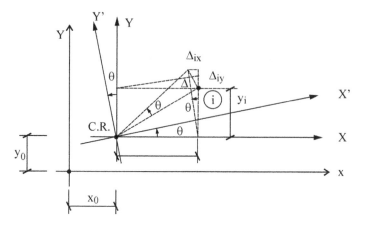

Figure 7.20 Displacement of the system due to effect of rotation

This means that the general twist generates simultaneously forces in both directions x and y, in each element of the system.

The moment of all these forces should balance the torsional moment M_t:

$$M_t = \sum_{i=1}^{j} (F_{ix}^{M_t} Y_i + F_{iy}^{M_t}) = \theta \cdot \sum_{i=1}^{j} k_{x,i} Y_i^2 + k_{y,i} X_i^2$$

$$\Rightarrow \quad \theta = \frac{M_t}{\displaystyle\sum_{i=1}^{j} k_{x,i} Y_i^2 + k_{y,i} X_i^2}$$

$$\Rightarrow \quad \begin{cases} F_{ix}^{M_t} = \dfrac{k_{x,i} Y_i}{\displaystyle\sum_{i=1}^{j} k_{x,i} Y_i^2 + k_{y,i} X_i^2} \cdot M_t \\[4mm] F_{ix}^{M_t} = \dfrac{k_{y,i} X_i}{\displaystyle\sum_{i=1}^{j} k_{x,i} Y_i^2 + k_{y,i} X_i^2} \cdot M_t \end{cases}$$

(7.10)

The total effect of translation and rotation is the sum of the two forces:

$$F_{ix} = F_{x,i} + F_{ix}^{M_t}$$

(7.11)

So, the total force, for each element, acting in direction x and, thereafter, y due to both effects – translation and rotation is determined.

Remarks

A.   The direct effect (translation) of the external force generates forces in walls parallel with its direction (for wall systems with parallel principal axes). In contrast, the torsional moment generates forces in walls in both directions.

B.   The total force produced by torsional moment, in each direction, is equal to zero, since no external force exists (when only the torsional moment acts) to balance them. Accordingly, the forces generated by the torsional moment have opposite sense within the system zones delimited by *X, CR, Y*.

C.   Accepting that the total effect of both translation and twist is the sum of the components generated by the two effects, it is implicitly assumed that their maximum effect occurs simultaneously. This assumption derives from that of *static action* of the external forces acting on the system. Actually, the seismic action generates coupled torsional and translation oscillations. According to the ratio between translation and torsional oscillation, the maximum response corresponding to two effects, (expressed in terms of kinematic parameters – displacements, velocities or accelerations as well in dynamic ones – forces or energies) can occur simultaneously or in different moments of the oscillation. Thus the accepted assumption is on the safe side.

7.6.5 Systems with non-parallel components

Although the case of system of walls with parallel principal axes is very often encountered, there are situations when this assumption is not fulfilled. This is the case, for instance, with buildings with irregular plane shapes or with buildings situated in block corner (Fig. 7.21).

   For such situations, the above derived notions and equations have to be generalized.

   The translation of the diaphragm in a direction induces, for each system component, displacements oblique in respect to its own principal axes. Consequently, wall stiffness about *oblique directions* (in respect to principal axes) has to be defined. The same notion, as well as the *principal axes*, has to be extended for the wall system considered as a whole (Fig. 7.22).

   Starting from these general notions, the direct effect (translation) as well as the torque effect on each component can be quantified.

a. Wall stiffness in respect to oblique axes

Let us consider a L-shape structural wall. Its cross-section has the two principal axes 1-1 and 2-2, inclined under the angle φ according to the rectangular axes *xx-yy* (Fig. 7.23).

   In order to compute the wall stiffness about *xx*, a displacement $\Delta_x = 1$ will be imposed. This displacement has two components about the directions 1-1 and 2-2, generating forces proportional to the corresponding stiffness. The sum of projections of these forces on axis *xx* is the wall stiffness about this axis:

$$k_x = k_1 \cos^2 \varphi + k_2 \sin^2 \varphi \tag{7.12}$$

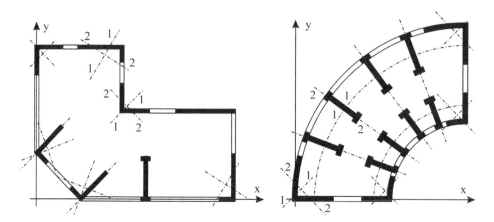

Figure 7.21 System with non-parallel principal axes walls

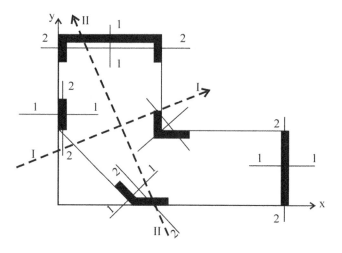

Figure 7.22 Coordinates of general wall system

Besides the force k_x, the displacement $\Delta_x = 1$ generates, simultaneously, a force on the direction yy

$$k_{xy} = (k_1 - k_2)\sin \varphi . \cos \varphi \tag{7.13}$$

Analog to the inertial moments, k_{xy} is called *centrifugal stiffness*.
Similarly, the stiffness corresponding to a displacement $\Delta_y = 1$ will be:

$$k_y = k + \sin^2 \varphi + k_2 \cos^2 \varphi \tag{7.14}$$

$$k_{yx} = (k_1 - k_2)\sin \varphi \cdot \cos \varphi \tag{7.15}$$

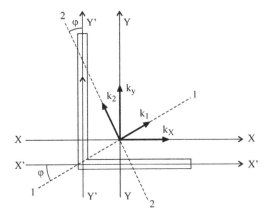

Figure 7.23 Axes of L-shape cross-section of a structural wall

b. Stiffness centroid coordinates of the system

Location of the stiffness centroid of the system (CR), according to the coordinate axes *xx-yy*, derives from the condition that the total stiffness about any axis *xx* or *yy* (the forces *k*) passes through CR. This is equivalent to the condition that the moment of all forces k_x, k_y, k_{xy} and k_{yx} about CR is zero. It results:

$$x_0 = \frac{\sum (k_y + k_{xy})x}{\sum (k_y + k_{xy})}$$

$$y_0 = \frac{\sum (k_x + k_{yx})y}{\sum (k_x + k_{yx})}$$

(7.16)

c. Principal axes of the system

Among all Cartesian axes that pass through the centroid CR, there is one that has the following attribute: a displacement of the system parallel to that generates forces only in this direction. Consequently, for the whole system,

$$\sum k_{xy} = 0 \quad (\text{respectively} \sum k_{yx} = 0).$$

(7.17)

These coordinate axes are called *principal axes of the system (I-I, II-II)*.

Total stiffness of the system according to the axes *I-I* and *II-II*, equal to the sum of the stiffness of all walls according to these axes, has maximum, respectively minimum, magnitude. Since, according to the definition, $\sum k_{xy} = 0$ the angle α between *I-I* and *xx* is given by (Fig. 7.24):

$$tg2\alpha = \frac{\sum (k_{1,i} - k_{2,i})\sin 2\varphi_i}{\sum (k_{1,i} - k_{2,i})\cos 2\varphi_i}$$

(7.18)

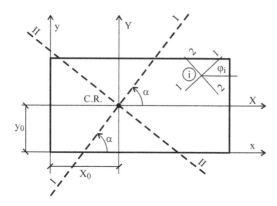

Figure 7.24 Principal axes of the wall system

d. Direct effect of horizontal force

Total horizontal force F has two components F_I and F_{II} according to the principal axes of the system *I-I* and *II-II*.

On their directions, these forces generate the displacements

$$\Delta_I = \frac{F_I}{\sum k_{I,i}}$$

$$\Delta_{II} = \frac{F_{II}}{\sum k_{II,i}}$$
(7.19)

where $k_{I,i}$ and $k_{II,i}$ are the wall stiffness on directions *I-I* and *II-II*. They can be determined with the formula for k_x and k_y by replacing the angle φ with α

For each wall, the displacements Δ_I and Δ_{II} have components about its principal axes *1-1, 2-2*:

- On direction *1-1* the displacements are $\Delta_I \cdot \cos \beta$ and $\Delta_{II} \cdot \sin \beta$
- On direction *2-2* $-\Delta_I \cdot \sin \beta$ and $\Delta_{II} \cdot \cos \beta$

where $\beta = \varphi - \alpha$ is the angle between wall axe *1-1* and the system axe *I-I*.

These displacements generate, in each wall, according to its principal axes a. and *2-2*, the forces:

$$F_{1,i} = \Delta_I \cdot \cos \beta + \Delta_{II} \cdot \sin \beta$$
(7.20)

$$F_{2,i} = \Delta_I \cdot \sin \beta + \Delta_{II} \cdot \cos \beta$$
(7.21)

These forces are the *direct effect* (translation) of the external force F on each wall i.

e. Effect of global torsion (twist)

For determining the forces generated by the torsional moment M_t in the wall i, according to the principal axes *1-1* and *2-2* the same approach as for walls with parallel principal axes will be used. It results:

$$F_{1,i}^t = \frac{k_{1,i} \cdot d_{1,i}}{\sum(k_{1,i} \cdot d_{1,i}^2 + k_{2,i} \cdot d_{2,i}^2)} M_t$$

$$F_{2,i}^t = \frac{k_{2,i} \cdot d_{2,i}}{\sum(k_{1,i} \cdot d_{1,i}^2 + k_{2,i} \cdot d_{2,i}^2)} M_t$$

(7.22)

The total force in wall i is the sum of forces resulting from the direct effect and from the torsional moment.

7.6.6 Special Analysis Issues

a. Additional Eccentricity

According to the elastic analysis of wall system it is assumed that *general twist* of the system occurs when eccentricity exists between the horizontal seismic force F and the stiffness centroid CR. The eccentricity is the distance between CR and the masses centroid CM, where it is supposed that the seismic (inertial) force is applied.

Actually, "parasitic" phenomena occur, which alter the theoretical magnitude of this eccentricity. Among these phenomena can be listed the following:

A. *Uncertainties regarding the real location of masses centroid CM.* At each building level, the true repartition of masses is not accurately known. So, the partitions' location can be changed during the building lifetime, supplementary masses can be added, some of initial existing masses can be removed, live load distribution is permanently changed, etc.

B. *Uncertainties regarding the real location of stiffness centroid CR.* Real stiffness of each wall can be only approximately assessed. Effect of differentiated cracking over the wall height, the real geometry of the wall cross section including the active width of flanges of T shape sections, real behavior of coupled walls, effectiveness of wall encasement at foundation level, and many other similar phenomena are ignored by current analysis.

C. *Effect of horizontal diaphragm flexibility* could change the contribution of different walls to the total stiffness of the system in comparison with that assessed according to the assumption of infinitely stiffness. Accordingly, the location of the axis of twisting can differ from the assumed one.

D. *Effect of progressive yielding of the walls.* During high intensity seismic actions, some walls experience incursions in post-elastic range of behavior. Accordingly, their stiffness is drastically modified in comparison with elastic stiffness and the location of CR is permanently modified.

In order to quantify globally such phenomena a supplementary eccentricity of the seismic force Δ_{e0} (additional to the geometrical distance between CR and CM) should be taken into account.

According to different structural codes, the additional eccentricity Δ_{e0} is considered to be proportional to the building length B or proportional to its in-plan dimensions A and B.

7.7 Design and Detailing of Cantilever Wall

7.7.1 General

Cantilever structural walls are subjected to lateral forces, determined through system analysis, and to longitudinal (vertical) forces that are the slabs' reactions.

Cantilever walls have specific structural peculiarities, which influence their analysis, design and detailing, as follows:

A. The *aspect ratio*, i.e. the ratio between wall length and its cross-section height, has magnitudes within a wide range influencing substantially the wall behavior. Thus, for *long walls* (aspect ratio over about five) the *flexural behavior* is predominant, while for *short (squad) walls* (aspect ratio under three) effect of *shear force* is predominant.

B. The cross-section shape can be various, with unusual proportions. For instance, they can have rectangular cross-section with a width normally of about 20 to 40 cm and a height of as much as several meters.

According to the general philosophy of seismic design, plastic deformations are expected during high intensity earthquakes. For cantilever walls, plastic deformations are located at the wall base, where the internal forces have maximum magnitude. They are developed along the *plastic length* l_p called also *potential plastic zone*.

Through specific design and detailing measures, it is ensured that the remainder element zone behaves elastically, even during the most severe earthquakes.

Long cantilever walls are modeled as linear members subjected to eccentric compression and shear force.

The internal forces are easily calculated since the element is statically determined. However, the *design forces* differ from those derived from equivalent seismic forces in order to control the development of plastic deformations. According to the *capacity design method* (see chapter 5.8) the design forces will be determined so that

A. Plastic deformations will develop only within plastic zone, keeping the rest of the element within elastic range of behavior;

B. The premature failure due to shear is prevented.

7.7.2 Flexural Design

The concrete cross-section is determined according to specific constructive rules (performing a *pre-proportioning*). Basic requirements are the following:

A. Average normal stress over the wall cross-section has to be limited in order to ensure ductile behavior: $\bar{\sigma} = \frac{N}{A_w} \leq (0.2-0.3)f_c$ (A_w is the wall cross-section gross area and f_c is the concrete compressive strength)

B. Average tangential stress over the wall cross section shall be limited for preventing brittle failure due to shear

C. Minimum thickness of 150 mm has to be observed in order to ensure the concrete casting.

The external forces on the wall are the *vertical loads* due to slab reactions and *horizontal forces* due to wind or seismic actions. In each section of the wall they generate axial forces, bending moments and shear forces.

The design rules differ for the two zones of the wall: that where plastic deformations are expected (*potential plastic zone*) and that with elastic behavior (the rest of the element).

The plastic potential zone developed at the wall base has a length provided by each code. As an example plastic length l_p can be accepted as much as $l_p = 0.4h + 0.05H$ (h is the wall width and H is its total heigh) rounded off to a multiple of story height.

Vertical loads at each wall level are the slab reactions. They depend upon the shape and proportion of the slab system (slabs directly supported by the wall or slab-beam systems spanning between walls).

Frequently, the walls have T, I, or similar cross-sections. The active flange width is determined through relationships like:

$$b_{eff} = b + \Delta b_1 + \Delta b_2 \tag{7.23}$$

where Δb_1 and Δb_2 are given by codes through empirical rules or relationships.

According to each specific code, Δ_b is determined differently for specific situations:

A. For calculating the *resistance capacity*. The active flange width differs for compressed and tensioned zone. The active flange calculated for resistance capacity is used for determining the *tributary areas* of the slab too.
B. For determining the wall *sectional stiffness*.

Horizontal loads on each wall result from *system analysis*.

Internal forces (bending moments, shear forces and axial forces), at each wall level, derive from equilibrium conditions since the element is statically determined.

Within potential plastic zone the bending moment is directly determined from the *equivalent seismic forces* acting on considered wall.

Design bending moments over the rest of the wall should have highest predictable magnitude in order to ensure elastic behavior of this portion of the wall. According to the *capacity design method* the design bending moments within elastic zone should be associated to the *maximum* moment capacity of the wall base. Thus, within elastic zone the bending moment will be:

$$M_{B,d} = M_B \cdot k_M \cdot \omega \tag{7.24}$$

where: M_B is the bending moment resulted from equivalent seismic forces

k_M – correction coefficient that includes the effect of over-strength as well as other potential amplification effects (dynamic, spatial, etc).

ω – ratio between moment capacity of the wall base section and bending moment (at same section) determined with code equivalent seismic forces.

Flexural behavior of the wall is ductile, semi-ductile or brittle depending upon the compressed zone depth x.

For $x < x_b$ the element failure starts with yielding of the tensioned reinforcement and is completed by crushing of the compressed concrete. Thus, large post-elastic deformations are expected and the element behavior is *ductile*.

For $x > x_b$ the failure is due to compressed concrete without yielding of the tensioned (or less compressed) reinforcement. Accordingly, the wall behavior is brittle (or, in any case, significantly less ductile than in the previous case).

Within plastic zone ductile behavior is required. Thus, the wall cross section should be proportioned so that the condition $x < x_b$ is fulfilled. For walls with rectangular cross section it is recommended to strengthen the boundary. So, a kind of "bulbs" are obtained at the wall edges that ensure not only a robust compressed zone but, also, impede the buckling of the wall edges, facilitate reinforcing bars arrangement and concrete casting. In special situations, when $x < x_b$ cannot be fulfilled, confinement of the compressed zone with close spaced hoops has to be implemented.

Wall design for *ultimate limit state* requires, generally, specific computer programs, since the cross sections have special shape and constructive reinforcement is provided along the wall face.

The resulting longitudinal reinforcement bars are provided toward the wall edges.

The bars are confined with closed stirrups similar to the case of columns.

These stirrups stabilize the longitudinal bars (impede their buckling) but don't participate in the shear resistance of the wall. In certain situations they are provided for concrete confinement.

The reinforcement at wall edge shall be over the *minimum amount* related to the local concrete zone (like for a local column).

Besides the calculated reinforcing bars, constructive reinforcement is provided along the wall sides. The amount of this reinforcement depends upon the geometry of the wall, the seismic zone and other criteria defined by codes. This constructive reinforcement participates to the wall flexural capacity.

7.7.3 Design for Shear

The shear force induces principal stresses that generate inclined cracks.

In order to prevent premature brittle failure due to shear, conservative approach for shear design is provided by codes.

Accordingly,

a. The magnitude of *design shear force* is the maximum that can be developed by the element
b. The *shear capacity* is assessed through conservative formula.

The design shear force is that associated to the yielding of the wall base (through bending), taking into account the over-strength and other adverse phenomena:

$$V_{Ed} = k_V \cdot \omega \cdot V_s \tag{7.25}$$

where: k_V is an amplification factor similar to k_M and

$$\omega = \frac{M_R}{M_s} \tag{7.26}$$

The shear/flexural failure occurs through an inclined critical crack. The "external" (in respect with the crack) shear force is balanced by the transverse reinforcement, which sews the crack, and by the concrete shear capacity V_c. The concrete capacity includes, actually, three effects: (1) the compressed concrete shear resistance (2) the aggregate interlocks contribution and (3) the dowel effect of the longitudinal bars.

One has:

$$V_{Ed} \leq V_R = \sum A_{stirr} f_s + V_c \tag{7.27}$$

Within plastic zone conservative magnitude of V_c is accepted (sometimes zero) so that a great amount of shear is resisted by the transverse reinforcement. It is also required to limit V_{Ed} in order to avoid excessive inclined crack width.

For resisting the shear forces, horizontal bars are provided on both wall sides, calculated according to the above-described procedure.

Within the remaining wall portion no plastic deformation is expected. Accordingly, the rules for transverse reinforcement design and detailing are less severe.

Connectors within construction (casting) joints

Besides the inclined cracks, the shear force generates a tendency of slip along horizontal sections where the element continuity is weakened due to interruption of continuous concrete casting.

According to the construction procedure, joints are provided, normally, at the top face of the slabs. Sometimes, construction joints are provided at the sections just under slabs. Even when taking proper measures to resume correctly the concrete casting, the discontinuity due to concrete of different ages constitutes a weakened section, where potential slip of adjacent wall portions can occur. In order to prevent this phenomenon, the construction joints have to be sewed by connectors.

Total connectors' area can be determined through *shear friction procedure*. According to this procedure the slip along the joint is prevented through friction forces generated by the axial force (normal to the joint) and by the clamping force developed by the connectors. The sliding *resistance L_R* over a joint crossed by connectors having the total area A_c determined by formula like:

$$L_R = \mu_f(A_s f_s + N) \tag{7.28}$$

where: μ_f is the frictional coefficient (concrete on concrete) estimated at about $\mu_f = 0.7-1.4$ (average value $\mu_f = 1.0$),

N – axial force on the joint and

f_s – connectors' strength.

The total area of connectors crossing the joint derives from the condition that the resistance L_R should be greater than the effective shear force acting along the joint.

7.7.4 Stiffness

According to the simplified analysis, the notion of member stiffness is suitably adapted.

The simplified analysis method assumes that, although each element is loaded with a set of lateral forces, the displacement compatibility is considered only at reference level located at building top or at 0.8 of its height. Consequently, the definition of element stiffness is "adjusted" and considered to be *the total lateral force ΣF_i that corresponds to a lateral displacement $\Delta = 1$ at the reference level* ($\Delta = \Delta_{top}$ *or* $\Delta = \Delta_{0.8H}$).

Thus, for determining the wall stiffness, the following steps have to be performed:

- Load the wall with a system of lateral forces F_i distributed similar to external horizontal forces (wind or seismic) i.e. uniformly or triangularly.
- Calculate the lateral displacement Δ at reference level (cantilever top or 0.8H)
- Determine the element stiffness

$$k = \frac{\sum F_i}{\Delta} \tag{7.29}$$

Provided that the same rule is accepted for all structural components of the system, the value k is a reasonable criterion for expressing the stiffness of each element.

7.7.5 Detailing

Constructive rules are specific to each structural code.

It is normally accepted that minimum wall thickness shall be: $b \geq 150\,$mm and $b \geq H_s/20$. The cross section of rectangular boundary elements of the wall will observe the following supplementary requirements: $h_p \geq 250\,$mm and $b_p \geq 2b$.

The wall flanges will have a minimum thickness of 150 mm and width of at least $H_s/4$.

The following types of reinforcements are provided within structural walls:

A. *Principal reinforcement* results from structural design and are:

1. Longitudinal (vertical) rebar that ensure the wall flexural capacity
2. Vertical bars provided along the wall faces
3. Transverse reinforcement (horizontal) which resists shear forces
4. Connectors crossing the casting joints
5. Hoops for confining the wall compressed zone (when necessary)
6. Stabilizing stirrups for preventing buckling of longitudinal bars.

B. *Constructive reinforcement* – is provided for resistance to internal forces that have not been explicitly taken into account within analysis and design process as, for example, those generated by concrete shrinkage, creep and differential temperatures.

Generally, the constructive reinforcement consists of two layers of welded wire meshes ϕ 5 mm at 200 mm spacing, provided on both faces of the wall.

Local reinforcement observes, generally, the rules deduced from Figure 7.25.

Maximum stirrups spacing is:

- *within potential plastic zone:*

    – 150 mm in seismic zone of moderate intensity;
    – 125 mm in seismic zone of high intensity but no more than 10d.

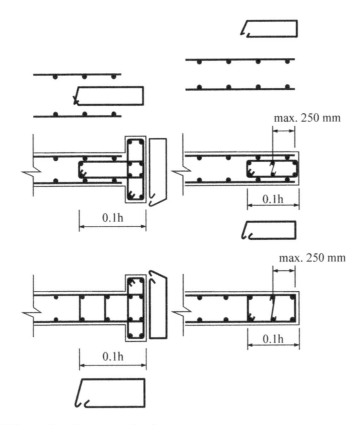

Figure 7.25 Local reinforcement of walls

- *within elastic portion of the wall:*

  – 200 mm, but no more than 15d.

- *Confining reinforcement of compressed zones*

When compressed zone depth x is over xb $(x > xb)$, confining reinforcement has to be provided within plastic potential zones, in order to enhance the wall ductility.

7.8 Coupled Wall Design and Detailing

Coupled wall consists of walls and coupling beams.

The walls are designed and detailed similar to the cantilever walls (see sub-chapter 7.7), for internal forces determined according to the previous paragraph.

Coupling beams design involves the following steps:

A. Determination of coupling beam thickness (width)
B. Flexural design that determines the longitudinal reinforcement area As
C. Design for shear.

Coupling beam thickness is determined analytically or through constructive rules. It can be the same as the walls or greater.

The most significant requirement for determining coupling beam thickness is to prevent excessive cracking due to shear. Codes quantify this requirement through inequality $V_{Ed} < 2bhf_{ct}$. When the inequality is not fulfilled for the beam with same thickness as the wall, the following solutions can be adopted:

A. Increase the beam thickness.
B. Diminish the bending moments at beam-ends up to 20% of those resulting from elastic analysis. Accordingly, the design shear forces (associated to the flexural capacity of the beam) have smaller magnitude.
C. Provide diagonal reinforcement.

Flexural design raises no special problems. Since the cross section is symmetrically reinforced, internal lever arm is the distance between longitudinal bars: $z = d - d_1$ and the necessary area of longitudinal reinforcement will be:

$$A_{s,nec} = M/f_s \cdot z \tag{7.30}$$

Effective reinforcement area has to be as close as possible to the necessary one, otherwise the flexural over-capacity induces shear forces above those considered in shear design.

Design for shear should be conservative in order to prevent premature (brittle) failure through shear without proper development of ductile flexural deformation. Accordingly,

A. *Design shear force* V_d will be greatest magnitude of shear that could be developed by the element and
B. *Shear capacity* will be computed through conservative reliable procedures.

Design shear force that fulfills the above requirement is that associated with the yielding of longitudinal bars, taking into account their potential over-strength:

$$V_{Ed} = \frac{|M_R^l| + |M_R^r|}{l} \tag{7.31}$$

where: M_R^l and M_R^r are the flexural capacities at the left- and right-end of the beam, computed with over-strength i.e. $M_R = A_s(1.25 f_y)z$ and l – the beam span.

Shear capacity will be computed accepting that the concrete (deteriorated by repeated cracking) doesn't participate in shear carrying. Thus the shear force is resisted only by transverse reinforcement. It is assumed that critical diagonal crack is inclined at an angle of 45°. Since the number of stirrups crossed by the diagonal critical crack is equal to (h_b/a_e) (h_b is the beam height and ae is the stirrups spacing), the shear capacity results:

$$V_R = f_{ywd}A_s(h_b/a_e) \tag{7.32}$$

where: A_s is the total area of a stirrup and f_{ywd} is the stirrup steel strength.

If A_s is chosen, from $V_{Ed} \le V_R$ the stirrup spacing results.

When design force magnitude is high (due to high magnitude of flexural capacity) the transverse reinforcement area can exceed the maximum amount allowed by codes or cannot be, practically, realized (it would require large stirrup diameter or excessively close spacing). In such cases, one of following solutions can be adopted:

A.   Decrease of flexural capacity, accepting plastic moment redistribution between beams and walls within a limit of 20% of elastic moments. Equilibrium between external and internal forces shall be observed within whole element as well as in each of its parts.
B.   Provide *diagonal reinforcement* besides the stirrups.

Coupling beams with diagonal reinforcement

Transfer of internal forces between beam ends through a truss mechanism, involving horizontal chords made by longitudinal reinforcement, inclined concrete struts and vertical "ties" made by stirrups, is difficult when the nominal intensity of tangential stress exceeds about $2f_{ctd}$ (f_{ctd} is the concrete tensile strength):

$$\bar{\tau} = \frac{V}{bd} \ge 2f_{ctd} \tag{7.33}$$

In such cases the concrete struts are deteriorated through repeated reversal forces. Moreover, the cracks developed in the vicinity of beam-ends don't close completely (due to residual plastic strain) and, to the moment reversal, can cut the whole beam section. Thus, the shear is carried throughout these sections only by the dowel effect of the longitudinal reinforcement and the shear capacity and stiffness drastically drop.

Providing diagonal reinforcement, these phenomena are fully prevented and a stable hysteretic behavior of the coupling beam is ensured.

Since the seismic shear force changes its sign, diagonal reinforcement should be provided in both directions.

Each diagonal reinforcement consists of a group of four or more bars tied together through spiral reinforcement aimed to prevent buckling of individual diagonal bars. Due to shear force reversal, diagonal bars are either in tension or in compression over their full length.

In order to ensure concrete casting and to avoid reinforcement agglomeration, the beam thickness (width) should be of about 280 to 300 mm (thickness of minimum 250 mm is required).

It is recommended to transfer whole magnitude of shear force through diagonal reinforcement, although constructive longitudinal bars and stirrups are provided too. Thus, the total area of diagonal bars of a group, inclined by an angle α, will be determined through the equation:

$$A_d = \frac{V_{Ed}}{2f_{yd} \sin \alpha} \tag{7.34}$$

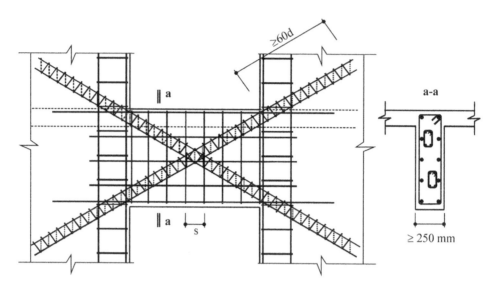

Figure 7.26 Coupling beam with diagonal reinforcement

Proper anchorage of diagonal bars has to be ensured, since they are subjected to high magnitude reversal forces and a tendency of progressive deterioration of the anchorage was evidenced.

Walls Coupling through Slab

An important advantage specific to wall systems is the possibility to use flat slabs without beams as floor systems. The solution can be implemented provided that the whole horizontal force is carried by the system of vertical walls (solid or with openings). Accordingly, in contrast with frame systems, no plastic deformations are necessary within the floor system for ensuring seismic energy dissipation.

Flat slabs (without beams) are easily executable from the point of view formwork and reinforcement, accommodate easily the equipment tubing, lead to diminishing of story height (in comparison with slab-beam systems) and, generally, lead to significant savings in material and labor.

Two issues have to be addressed when flat slabs are implemented in connection with structural wall system:

1. Analysis, design and detailing of slab for gravity loads and
2. Specific verifications related to slab seismic response.

1. Due to unusual slab pattern the load transfer to bearing elements (walls) cannot be easily anticipated. So, accurate analysis using elastic finite element software has to be performed. Both internal forces (bending moments and shear) and deflections have to be determined. It will be found that short slab strips, connecting closely located walls, act as bearing elements (similar to beams) for the neighboring slab panels, developing

important bending moments and sometimes substantial shear force. The slab height will be determined so that it complies with the following requirements:

- Flexural reinforcement amount should be within economic domain (reinforcement percentage of about 1–1.2% in zones with maximum moment).
- No shear reinforcement is required (concrete shall be able to carry the whole shear force)
- Deflections have to be within admissible limits under service loads
- Slab height of 170 mm to 250 mm (sometimes 300 mm) is currently required.

2. Slab seismic response should remain within elastic behavior domain. For checking this requirement the magnitude of slab internal forces generated by the seismic action considered with its maximum magnitude should be determined. (Actually, the structural response under equivalent seismic forces will be multiplied with appropriate factor). High magnitude moments and shear forces in the short slab strips will again be found. Reinforcement similar to that of beams (rebars tied with stirrups) can be added in such a situation if necessary. Another way to diminish the local concentration of efforts is by providing spacing between walls over about 1.50 m.

7.9 Diaphragms

Systems with structural walls involve, generally, non-uniform in-plan distribution of vertical bearing elements (walls). Sometimes large spacing between them exists.

Due to these peculiarities of wall systems (in comparison with framed structures) the role of horizontal diaphragms, as elements that coordinate the overall seismic response, is particularly important and, sometimes, difficult to fulfil.

Special attention has to be paid to the following situations:

a. Prefabricated floor systems
b. Floors with large openings for staircases, elevators, etc.
c. Large distances between walls
d. Systems with non-uniform vertical distribution of the walls, between floors (i.e. walls interrupted at certain stories)
e. Buildings with large in-plane irregularities.

The following aspects should be checked within analysis and design process:

- Diaphragm resistance and stiffness
- Connection between diaphragm and walls.

As a general rule, diaphragms have to respond elastically to the seismic action. Accordingly, the highest magnitude of forces to be transferred between vertical elements has to be assessed. They will be the "external" design loads for the diaphragms.

According to an "exact" approach, the analysis of diaphragm should be performed together and in interaction with the wall system. Procedures based on finite element method (FEM) of the whole structural system should be applied.

A simplified procedure for assessment of internal forces due to seismic action within a horizontal diaphragm, accepted by codes, considers it a horizontal beam supported

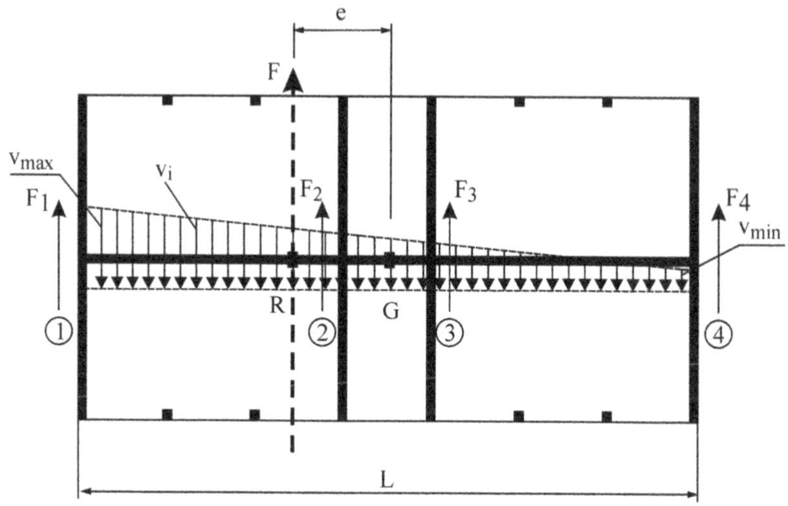

Figure 7.27 Simplified distribution of seismic forces over diaphragm

by the walls. Assuming that the floor mass is uniformly distributed, and the floor has a rigid-body translation and a rotational displacement due to seismic action, the inertial seismic forces, acting on each story, are distributed as shown in Figure 7.27. They are balanced by seismic forces on each wall that act as reactions of the horizontal beam (i.e. the diaphragm).

So, the diaphragm becomes a continuous horizontal beam subjected to the distributed inertial seismic forces and to the support reactions, which are the walls' forces. Bending moments and shear forces in each section of diaphragm can be subsequently determined. Depending upon the ratio span length/beam height the diaphragm is either a *deep beam* or a *long* (ordinary) *beam*. Accordingly, the flexural and the shear reinforcement will be determined.

Besides the diaphragm design as beam with bending and shear, connection with the vertical walls has to be ensured. For this purpose, shear friction approach is used.

7.10 Infrastructures and Foundations

7.10.1 *General*

In comparison with frame structures, wall systems show specific peculiarities that strongly influence the infrastructure conception and design, namely:

a. in-plane significant non-uniform distribution of strength and stiffness of vertical resisting elements (cantilever or coupled walls);
b. sometimes, great distance between vertical elements;
c. high magnitude of the flexural capacity at the base section of the walls.

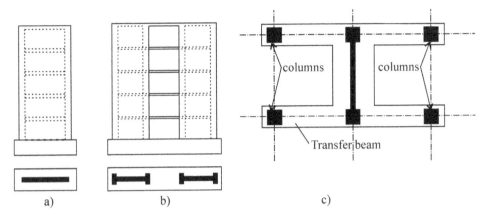

Figure 7.28 Individual foundations for structural walls: a) individual footing for single wall, b) Footing under two walls, c) Footing with transfer beam

So, the foundation system of structural wall buildings have to transfer to the soil high magnitude bending moments and shear forces, non-uniformly distributed in plane, and, sometimes, spaced at significant distances.

Basically, following foundation systems could be implemented:

1. *Individual footings* for each wall or for groups of walls. This is the case of buildings of low or moderate height, without a basement.
2. *Infrastructure provided under the whole building.* This is the case of buildings with basement extended over one or more levels.
3. *Special foundation* systems: pile foundations, rocking wall systems, etc.

7.10.2 Individual Foundations

Structural walls subjected to moderate overturning moments can be provided with individual footings.

According to the general layout of the building and to the magnitude of forces to be transferred to the soil different solutions of the foundation system can be adopted as, for example (Fig. 7.28):

A. Individual footing for single wall (Fig. 7.28a)
B. Footing under groups of two walls (Fig. 7.28b)
C. Footing with transfer beams or walls (Fig. 7.28c).

The main issue raised by such a solution is the fact that, at the wall base, high magnitude overturning moment and relatively small axial force act. In order to impede any lifting tendency of the foundation from the soil, solutions have to be found to expand the bearing area on the soil and collect axial force from neighboring columns. Providing transfer walls at the basement level, which connect the footing with neighboring elements, is a potential solution to fulfill this requirement.

During high intensity seismic actions, foundation and soil have to be kept within elastic range of behavior. The dissipating mechanism should be provided within super-structure. According to the *method of capacity design*, these requirements could be achieved if the infrastructure and the foundations are designed to respond elastically to the highest magnitude of forces transmitted by the superstructure. So, the design forces acting on the infrastructure and foundations will be:

A. flexural capacity with over-strength of wall base
B. shear and axial force associated to the wall flexural capacity (with over-strength)
C. self-weight of the foundation system.

7.10.3 Infrastructures

Wall systems of medium- and high-rise buildings require foundation solutions expanded under whole area of the building. These are foundation systems with *infrastructures*.

As already stated the infrastructure is the component of the structural system that shows, in comparison with the superstructure, a considerably greater *stiffness* and *resistance* for lateral forces.

The infrastructure is obtained through addition of supplementary structural walls at the basement level. Anyway, buildings provided with basement require *peripheral walls* along basement perimeter. These walls act, primarily, as retaining walls. Simultaneously, they lead to a sudden increase of stiffness and resistance of the structural system.

Due to the change in stiffness at the infrastructure level (in comparison with stiffness distribution above this level) a redistribution of internal forces, carried by different components, occurs. The total lateral force acting at walls' bottom (above the infrastructure) is shared to peripheral walls and internal walls, proportionally to their stiffness. The diaphragm (i.e. slab system just above the infrastructure) ensures the transfer of total lateral force to the infrastructure components. The wall system at basement level interacting with the horizontal diaphragms (slabs) and with the raft behaves similar to *multi-box system*.

Even though the analysis has been performed as accurately as possible (using FEM approach) certain redistributions between components due to interaction with the soil should be expected. So, the magnitude of shear force in certain internal walls and, accordingly, the bending moments diagram could be modified (diminished) due to the slide and rotation of the foundation on the soil. This redistribution is not easy to accurately quantify. Consequently, more assumptions concerning the degree of foundation embedment into the soil have to be accepted and the most reliable value of internal forces selected.

Design and detailing of infrastructure components. The following components have to be examined:

*   *Horizontal diaphragms*
*   *Interface diaphragm/walls.* At the slab/wall interface connectors have to be provided dimensioned according to the *shear friction* procedure

- *Walls*. Currently the infrastructure walls have to carry substantial shear forces being shear-dominated elements. Conservative rules for shear design of infrastructure walls should be observed as, for example, to neglect the contribution of the concrete to shear resistance of the wall.
- *Wall foundations* or *raft*.

As a general rule, robust detailing solutions have to be chosen since the analysis involves many uncertainties.

7.11 Case Study

7.11.1 General

Case study is about *a 19-story residential building with two-level basement*.

The building is located in Bucharest/Romania. It is included in a residential assembly involving two 14-story units and a 19-story one. The three buildings are separated through wide seismic joints. The basement, with garage function, is extended over the whole assembly. No joints were provided at basement level.

Designers:
ALL PLAN CONSTRUCTION s.r.l. Bucharest
Architects: E. Padure & R. Tempea
Structure: L. Crainic & A. Zybaczynski
Infrastructure: A. Marcu

7.11.2 Conceptual Design

The initial and most important task was to conceive the structural system which accommodates the architectural layout and complies with functional requirements at basement level where a garage function had to be ensured.

The first attempt was to adopt a dual system with columns and minimum number of structural walls. This solution would offer maximum freedom in using basement space for a garage. Due to the great magnitude of seismic lateral force and significant construction height (54 m) structural components resulted with unacceptably large sizes. Consequently, attempts have been made to convert more partitions into structural walls. Limitations were at the basement level since difficulties occurred in ensuring car access. Step by step, a solution has been found for cars' sheltering and for their access to garage cells (Fig. 7.29b). It was accepted to use overlapped sheltering for each two cars, using appropriate equipment. The clear height required for basement stories was 4.00 m.

The resulting structural walls had initially complicated the cross section shape. For rectifying this disadvantage intersected walls were slitted and the resulted joint was filled with light concrete blocks (Fig. 7.31). The resulting structural walls are planar with oblong rectangular cross section. An exception was the elevators/staircase cage which was considered as a tubular section.

The foundation solution resulted from two important circumstances:

- Presence of basement having many strong structural walls and
- Pretty poor resistance of foundation soil.

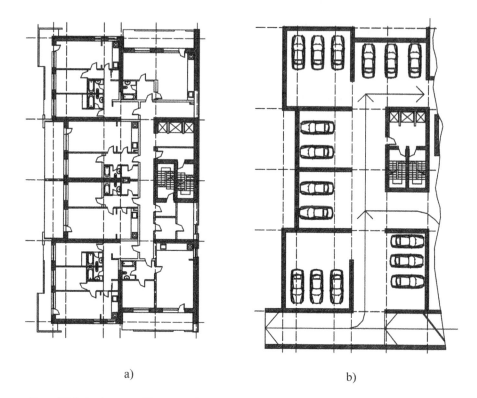

a)   b)

Figure 7.29 Architectural layout a) Current floor, b) Basement

Accordingly, the adopted foundation system had an infrastructure consisting of base-ment peripheral walls, internal structural walls, flat slabs at basement level and over basement as horizontal diaphragms, and a 1.80 m thickness raft. The infrastructure reactions due to gravity and seismic action are transferred to the foundation soil par-tially through the raft and partially through a system of bored reinforced concrete piles.

Initially the floor structure was a 14 cm slab supported by walls and beams. Fol-lowing the constructor's request, the initial slab-beam system was replaced by a 17 cm thick flat slab. At the basement level, due to high magnitude live loads, the slab is 20 cm thick and was provided with drop panels in the vicinity of structural wall edges (Fig. 7.32).

7.11.3 Preliminary Analysis and Design

Loadings (characteristic magnitudes):

1. Permanent actions (besides the self-weight of primary structure members which will be automatically determined by the structural analysis software): 2.0 (kN/m^2)
2. Variable actions (for residential building): 1.5 (kN/m^2)

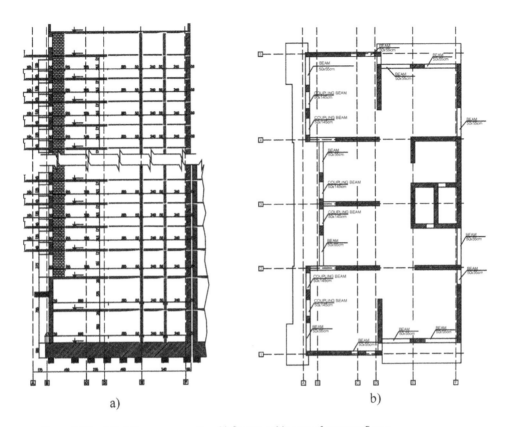

Figure 7.30 a) Building cross section. b) Structural layout of current floor

3. Snow (on roof): 2.0 (kN/m^2)
4. Variable actions on garage slabs: 5.0 (kN/m^2) (garage with two overlapped cars)

Seismic action corresponds to a Peak Ground Acceleration $PGA = 0.24$g. An importance factor of 1.2 has been considered.

Seismic equivalent total force is $F_b = 0.16 W_t$ (W_t is the total building weight)

Preliminary Seismic Design

Using simplified (half-empirical) formula and constructive rules, walls' thickness will be initially determined as follows:

a. For limiting the effect of shear force in walls, total walls' cross section area will fulfill the following inequality $F_b/A_w \leq 2f_{ctd}$ where A_w is the total area of walls parallel to the seismic force and f_{ctd} – tensile design strength of concrete.
b. For assuring ductile behavior each wall cross section will observe the following inequality: $n_0 = N/Af_{cd} \leq 0.3$.

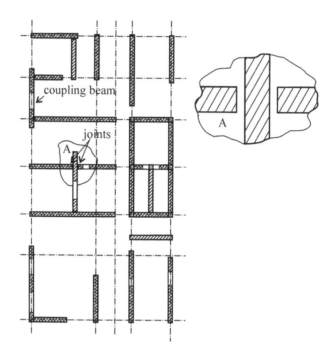

Figure 7.31 Slit of walls with complicated cross section

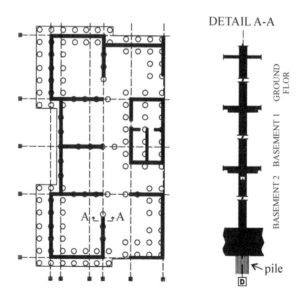

Figure 7.32 Basement layout with piles distribution and cross section

Table 7.1 Dimensionless axial force in walls

Wall #	Width (m)	Thickness (m)	Axial Force (kN)	n_0
T1	6.50	0.5	5100	0.105
T2	4.10	0.5	1610	0.052
T3	11.00	0.5	9842	0.119
T4	11.00	0.5	4352	0.053
T5	11.00	0.5	12100	0.147
T6	6.50	0.5	5510	0.113
T7	5.70	0.5	4850	0.113
L1	4.00	0.5	6000	0.200
L2	5.40	0.5	5600	0.138
L3	5.30	0.5	4350	0.109
L4	7.60	0.5	7000	0.123
L5	2.60	0.5	7400	0.379
L6	5.50	0.5	5000	0.121
L7	5.30	0.5	7135	0.179
L8	8.70	0.5	11460	0.176
L9	8.60	0.5	8970	0.139
L10	10.30	0.5	4755	0.062
L11	7.90	0.5	5600	0.095
NC	53.85	0.5	19500	0.048

Condition a. is fulfilled for walls' thickness of 500 mm. A check of condition b. is performed according to Table 7.1 for all walls.

In order to check if the dynamic configuration of the chosen structural system is advantageous a FEM model has been defined and its seismic response determined using an appropriate software package.

The first three vibration modes have following eigen periods:

Mode 1 T = 1.665 sec (translation)
Mode 2 T = 1.122 sec (translation)
Mode 3 T = 0.791 sec (torsion).

Consequently, the dynamic configuration of structural system was considered to be satisfactory.

Next step was to verify the fulfillment of drift control requirement. Due to the structure's high stiffness the requirement is met without any problem.

The general conclusion of these verifications is that the chosen structure does comply with global and local requirements.

7.11.4 Accurate Structural Analysis

Detailed analysis 3D model includes superstructure components (cantilever and coupled walls and horizontal diaphragms), infrastructure with its walls, diaphragms, and raft, the piles system and the foundation soil.

Walls, diaphragms and raft were modeled with elastic finite elements of shell types; foundation soil was considered as elastic Winkler's springs and piles were modeled with springs having elastic stiffness determined by analytical relationships and validated through several in situ tests.

Seismic action was quantified through design spectra provided by the code in force. Spatial effects of the seismic action and the influence of higher vibration modes have been taken into account through standard procedures.

Combinations of gravity and seismic actions have been considered according to the code provisions.

7.11.5 Design and Detailing

Structural components design and detailing was done according to the principles of capacity design method.

The dissipating mechanism was chosen with plastic deformations at the base of cantilever walls and within coupling beams.

Walls involve two zones: a potential plastic zone, at the wall base, (Zone "A") and the rest of the wall assumed to respond elastically to the lateral forces (Zone "B"). Zone A is designed to resist bending moment resulted from elastic analysis and to carry the shear force associated with maximum moment which can be developed by the critical plastic section which is the moment with over-strength. Zone B has to resist elastically to the maximum moments which can occur. Accordingly, the moments resulting from initial analysis will be magnified with appropriate factors which take into account the

Table 7.2 Design and detailing of a cantilever wall

Story	Zone	N [kN]	M_S [kNm]	V_S [kN]	M_d [kNm]	V_d [kN]	M_R [kNm]	$A_{w\,nec}/$ [mm²/m]	Effective horizontal web reinforcement
19	B	−148.28	488.212	135.95	692	203.925	7865	0	2Ø10/175
18	B	−552.27	642.138	30.5	910	45.75	8437	0	2Ø10/175
17	B	−957.53	783.671	106.31	1111	159.465	9294	0	2Ø10/175
16	B	−1364.3	600.211	221.32	851	331.98	9719	68	2Ø10/175
15	B	−1772.85	904.21	322.63	1282	483.945	10951	148	2Ø10/175
14	B	−2183.32	2008.414	407.48	2847	611.22	11992	295	2Ø10/175
13	B	−2595.77	3327.948	475.93	4718	713.895	12541	374	2Ø10/175
12	B	−3010.19	4820.71	530.16	6834	795.24	13500	418	2Ø10/175
11	B	−3426.48	6452.077	572.77	9146	859.155	31960	487	2Ø10/175
10	B	−3844.41	8197.754	607.34	11621	911.01	32805	532	2Ø10/175
9	B	−4263.66	10052.97	640.51	14251	960.765	33957	586	2Ø12/175
8	B	−4683.72	12052.99	685.78	17086	1028.67	34715	626	2Ø12/175
7	B	−5103.95	14309.72	768.89	20285	1153.335	35873	763	2Ø12/175
6	B	−5523.51	17064.63	933.21	24190	1399.815	36927	837	2Ø12/175
5	B	−5941.45	20753.42	1243.62	29420	1865.43	37572	945	2Ø12/175
4	B	−6356.75	26050.15	1779.4	36928	2669.1	38700	1083	2Ø12/175
3	A	−6768.83	34026.69	2672.07	37104	4008.105	48100	1177	2Ø12/175
2	A	−7249.23	45210.82	3970.45	49300	5955.675	49300	2250	2Ø16/175
1	A	−7943.39	34215.98	2987.26	41042	4480.89	43000	1248	2Ø14/150
−1	1	−1119.87	18673.81	4645.04	22399	6967.56	35200	2311	2Ø16/150
−2	1	−929.19	16665.16	4872.09	19990	7308.135	34000	2491	2Ø16/150

maximum lateral force which can be developed by the structural system and other adverse phenomena.

As an example, the design steps of one of the structural cantilevered walls are synthesized in Table 7.2. Wall thickness is 50 cm and the length 490 cm. For zone B the seismic elastic moments are multiplied with the factors ω and $K_M = 1.3$:

$$\omega = \frac{M_{Rd}}{M'_E} \tag{7.35}$$

where M_{Rd} is the seismic bending moment at considered section and M'_E – capable moment at the base section determined with over-strength and with effective reinforcement area.

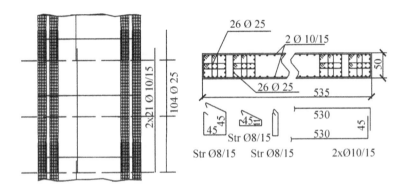

Figure 7.33 Detailing of a cantilever wall

Table 7.3 Coupling beam reinforcement design

Story	b [mm]	V [kN]	$A_{s,nec}$ [mm²]	Effective A_s
1	500	1487.92	4397	6Ø25
2	500	1007.00	2976	6Ø25
3	500	689.04	2036	6Ø25
4	500	584.93	1729	6Ø25
5	500	540.20	1596	6Ø25
6	500	508.87	1504	4Ø25
7	500	479.29	1416	4Ø25
8	500	449.69	1329	4Ø25
9	500	419.98	1241	4Ø25
10	400	389.85	1152	4Ø25
11	400	358.86	1061	4Ø25
12	400	326.49	965	4Ø25
13	400	292.05	863	4Ø25
14	400	255.01	754	4Ø20
15	400	215.84	638	4Ø20
16	400	193.32	571	4Ø20
17	400	167.79	496	4Ø20
18	400	124.60	368	4Ø20
19	400	128.44	903	4Ø20

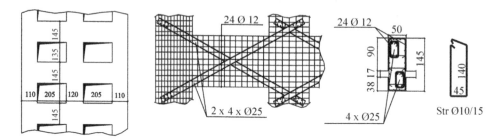

Figure 7.34 Detailing of a coupling beam

   Coupling beams are subjected to high intensity shear forces. Accordingly, diagonal reinforcement has been adopted. Constructive stirrups and longitudinal reinforcement (minimum amount) have been provided too. The design of a coupling beam 145 cm high and 205 cm in length is shown in Table 7.3. In Table 7.3, A_s refers to the diagonal reinforcement.

Conclusions

Due to their "natural" resistance and stiffness to lateral forces, the structural wall systems significantly extended the use of reinforced concrete to high- and super-high-rise buildings subjected to strong seismic actions. Within the present chapter the behavior, analysis, design and detailing of structural walls systems and of their components are thoroughly examined. A case study of a typical wall system seismic prone building is presented.

Chapter 8

Dual Systems

Abstract

Frames associated with structural walls constitute a wide family of structures called dual systems. They combine the advantages of both types of structural systems: frames and structural walls being able to accommodate various architectural needs. Consequently, dual systems are nowadays perhaps most frequently used structural solution for high- and super-high-rise seismic prone buildings. Resulting from a combination of two types of elements – walls and frames – having very different deformation patterns under lateral loads, specific behavior features of dual systems are highlighted. Analysis, design and detailing approach of dual systems are described. Solutions for their infrastructures and foundations are presented. Two case studies presenting in detail the design and detailing steps are included.

8.1 General Considerations

Both pure frame structures and pure walls systems show specific disadvantages which limit their usage domain. If frames are associated with structural walls a wide family of structures results which combine the advantages of both types of structural systems: frames and structural wall systems. They are dual systems.

By implementing advantageously structural walls within a framed-type structure, structural requirements related with lateral force resistance and stiffness can be easily fulfilled. The system accommodates various architectural needs including those related with special shape of modern buildings (Fig. 8.1).

Consequently, dual systems constitute optimum solution for multistory high-rise office buildings and for hospitals, luxury hotels, schools, etc.

According to the contribution of two main components – frames and walls – to the shear resistance, codes classify the dual systems in two main types:

- *frame-equivalent dual system* in which the shear resistance at system base is provided mainly (more than 50%) by the frames and
- *wall-equivalent dual system* for which the contribution of walls to the base system shear resistance is higher than 50%.

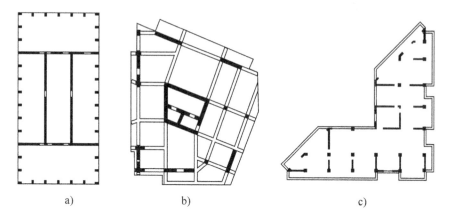

a)   b)   c)

Figure 8.1 Dual structural systems for various architectural layouts

8.2 Behavior of Dual Systems

Although the dual systems are very frequently used as optimal solution for a wide range of multistory buildings, the laboratory tests (static, pseudo-dynamic or those performed on shaking table) about their seismic behavior are scarce (Paulay, T. & Priestley, M.J.N. 1992). More information is obtained through advanced seismic analysis using up-to-date powerful software packages or by processing lessons drawn from recent major earthquakes.

Most important specific behavior features of dual systems is a result of combination of two types of elements – walls and frames – having very different deformation patterns under lateral loads.

Basically, the frame deformation under lateral loads is of shear type. Cantilever structural walls show a flexural deformed shape. In a dual system the two types of elements are inter-connected by the horizontal diaphragms (slabs) so that their deformations will be forced to be the same. Literature currently presents as typical case that of a dual system substructure consisting of a combination of shear-type frame with cantilever wall (Fig. 8.2).

The following behavior features of this structure can be deduced from these drawings:

- Deformed shape of the independent frame under lateral loads is of shear type (Fig. 8.2a) while that of the wall is flexural (Fig. 8.2b)
- The horizontal diaphragms (modeled as infinitely stiff pinned links) impose on the two components the same lateral displacements at each story. Due to different tendencies to deform of the two components, the diaphragm develops link forces – tension at system lower part and compression at its upper part. Accordingly, the wall upper part contribution to resist lateral forces is diminished while that of the frame is increased. Towards its base the wall is a stiffer element with higher contribution to carrying the total shear while the frames' contribution is lower.

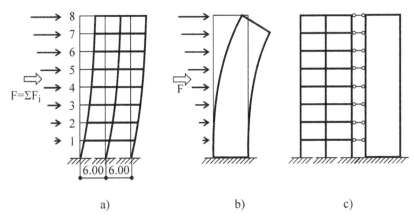

Figure 8.2 Typical combination of dual system components a) Frame system, b) Wall system, c) Assembly system

- Horizontal diaphragms of dual systems are subjected to important internal forces (in comparison with those of pure frame or pure wall systems) due to the inter-action between two kinds of components – frames and walls – and due to great distances between stiff components. These forces can generate large opened cracks evident in diaphragms situated at the lower part of the building (diaphragms subjected to high tension).

In fact, the real dual systems are large variety of spatial combinations of different types of structural walls (cantilever, coupled walls) each of them with its specific behavior.

As an example, a frame like that of Figure 8.2a with two equal spans of 6.00 m and story height of 3.50 m, having column sizes of 600 mm × 600 mm (external) and 700 mm × 700 mm (internal) and beams of 300 mm × 650 mm shows a shear-flexural behavior similar to that of the frame from Figure 6.1c (see Chapter 6). Interacting with a cantilever wall of 6.00 m width and of 300 mm thickness the deformed shape under lateral forces of two components acting individually and inter-connected is shown in Figure 8.3a. The forces in links vary as shown in Figure 8.3b. Note that only the link at the structure top is compressed, all others being tensioned.

If a pushover analysis is performed for the assembly, dramatic changes in distribution of forces in links occur (Fig. 8.4b). They are generated by the plastic hinges developed in wall's components – beams and columns – which, step by step, modify the relative stiffness of two components: frame and wall.

Important differences between overall system behavior, as compared with that of the very simple plane model above examined, occur even though the qualitative phenomena listed herein remain valid. These differences result from: (a) spatial behavior of the system which implies torsional response, (b) gradual plasticization of system components, (c) foundation rotation because of soil deformability, (d) dynamic character of real response.

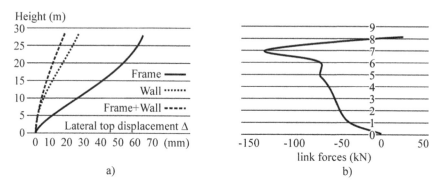

Figure 8.3 Elastic deformation shape a) two individual components and b) interconnected

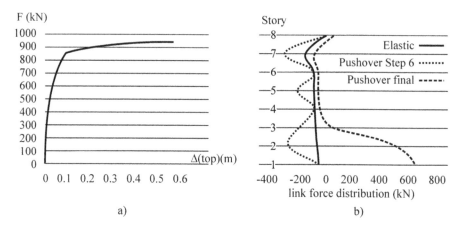

Figure 8.4 Pushover analysis of frame-wall assembly a) F–Δ curve, b) Link force distribution at two pushover steps as compared with elastic distribution

8.3   Conceptual Design of Dual Systems

The rules for system good conformation detailed within chapter 6 (frames) and 7 (walls) remain valid for dual systems with some adaptations.

It has to be pointed out that the structural designer of dual systems has many possibilities to accommodate the solution with various architectural and functional requirements in order to obtain optimal seismic response. In respect to this feature some basic rules have to be strictly observed since the irregular distribution of walls and frames, specific for dual systems, can easily lead to unbalanced structure. These basic rules can be listed as follows:

* Dominant vibration modes have to be translation; torsional vibration has to be limited. For example, dual system with an opened central core without torsional strength reserves shall be avoided, even if from the elastic analysis point of view

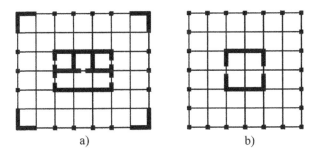

Figure 8.5 Favorable and unfavorable dual systems a) Favorable dual system, b) Unfavorable dual system

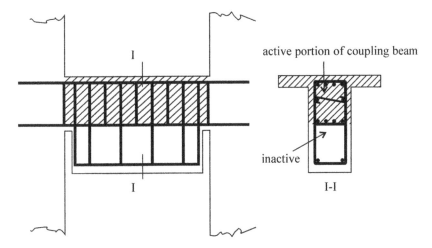

Figure 8.6 Partially slit of a too rigid coupling beam

the system is acceptable (Fig. 8.5b). Systems with walls on both directions situated on building contour are very stable for torsional effect (Fig. 8.5a). When more walls are provided along the building periphery failure of one wall doesn't reduce significantly the system torsional stiffness.

It is important to check the fulfillment of this requirement from the early stages of the design performing accurate structural analysis with reliable computer software. The in-plan distribution of components' stiffness – especially walls – could lead to unfavorable torsional effects. When necessary slit large walls (too rigid) through seismic joints or partially slit coupling beams (modify resistance and stiffness) in order to obtain a desired overall response (Fig. 8.6).

- Fulfillment of drift control requirement has to be verified at all building levels
- Vertical structural components – walls and columns – have to be kept continuous over the whole building height up to the foundation.

- Rigid and strong infrastructure is particularly recommended for dual systems which involve very non-uniform in-plan distribution of structural components (walls and frames) with different stiffness.

Each change in initial structural configuration should be checked through analysis until the fundamental response requirements are optimally met.

It has to be noted that the share of forces between elements and their distribution within each component can be very unusual in comparison with uniform regular systems.

8.4 Analysis, Design and Detailing of Dual Systems

Due to important in-plan non-uniformity currently encountered with dual systems accurate analysis involving the whole structural system (superstructure, infrastructure and foundation soil) is required. Simplified procedures are not applicable and can generate substantial errors.

Because of their complexity, the analysis of dual systems is currently performed using elastic approach with seismic action modeled through equivalent forces provided by code. For wall dominated systems finite element method is preferred to model cantilever and coupled walls as well as horizontal diaphragms.

The differential cracking of structural components generates important redistributions of internal forces. This effect is taken into account by applying correction factors to concrete elasticity modulus according to dominant stresses of different elements: columns and piers eccentrically compressed or subjected to (eccentric) tension, flexural elements, shear dominated elements (like coupling beams). As guidance, the following reduction factors of concrete elasticity modulus can be adopted:

- Marginal compressed columns and piers: 0.50–0.60
- Central columns (compressed): 0.70–0.80
- Marginal tensioned columns and piers: 0.20–0.30
- Coupling beams: 0.20–0.40.

Lower values of these factors take into account (indirectly, without properly quantifying) potential moment redistribution due to plastic deformation.

Across the structural elements' section the stresses provided by the FEM software are integrated to obtain internal forces: M, V, N. They are used for design and detailing of structural members.

It should be noted that, within analysis and design process, the theoretical accuracy is sacrificed for the need to obtain practical results through current approaches and code provisions. So, in a first step, the FEM approach is used to determine stresses in members considered as 2D elements. Then the stresses are converted into internal forces and the members are designed as linear bars (1D element).

Post-elastic behavior can be (practically) quantified by modeling the coupling walls through *equivalent frame model* (see chapter 7). In this way moment redistribution between structural components can be operated up to 30% of elastically calculated moments. In such cases the overall structure equilibrium has to be strictly observed.

In order to fulfill the advantageous dissipating mechanism requirement at design level the steps of *capacity design method* are recommended (see chapters 6 and 7).

Accordingly, the plastic mechanism should be foreseen, by choosing the location of plastic hinges (or zones). Then the hierarchy of members' capacities is determined using the over-strength and correction factors as shown in chapters 6 and 7 and the design magnitudes of internal forces are obtained.

Plastic zones will be detailed according to the rules meant to ensure good ductility by preventing plastic buckling of longitudinal reinforcement, brittle failure due to excessive compression or longitudinal steel amount and shear brittle failure. Elastic zones are designed according to current design approach.

Rules for detailing are those for frames and structural walls. Special attention should be paid to the design and detailing of diaphragms. They should behave elastically under important in-plan forces generated by the fact that they have to transfer high magnitude reactions between elements of different types and be spaced at large distances. Good connection to the system elements is required. In respect to this requirement shear friction procedure and local "beam effect" should be observed and implemented.

8.5 Infrastructures and Foundations

Due to functional and esthetic considerations (need of large open spaces, special in-plan and elevation building shape, parking at the basement levels with easy cars' circulation) dual systems often involve unusual structural layout with pronounced irregularities. Vertical superstructure elements with high rigidity can be provided over large distances transferring to the foundation system high magnitude reactions. To cope with these specific features of dual systems, especially for medium and high-rise buildings exposed to strong seismic events, robust infrastructures are required. For this purpose, at the basement level, often extended over more levels, an additional (as compared with the superstructure) system of structural walls has to be provided. The structural walls advantageously arranged, interacting with the peripheral walls and with horizontal diaphragms (slabs) and with the raft, ensure a multi-box system behavior. Such a system shows a very high flexural and torsional stiffness and resistance, similar to a multi-box bridge deck. It is important to understand that the raft alone, considered as an isolated structural component, cannot, normally, transfer to the soil the reactions of the superstructure.

According to the capacity design principle, infrastructure should respond elastically to the maximum seismic reactions which can be developed by the superstructure. They are the flexural capacities of superstructure components (walls and columns) determined with over-strength, the associated shear forces and the vertical axial forces. Due to their in-plan non-uniformity, these forces, acting together with the soil reactions, generates to the infrastructure considered as a whole important bending moments in vertical plans, high magnitude shear forces and torsional moments about horizontal axes. Because of high intensity shear forces in infrastructure walls they can require large width cross sections which can be as much as 100 cm or even more. When necessary, piles can be provided for ensuring additional supports to the raft.

8.6 Case Studies

Case # 1: Residential building with 19 stories and two level basement.

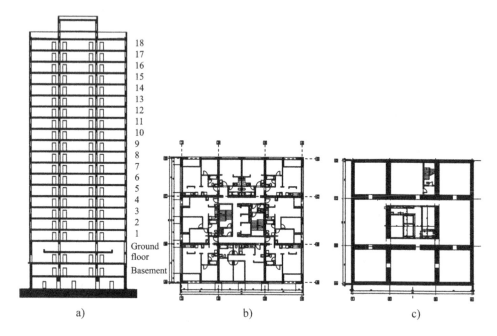

Figure 8.7 Architectural layout a) Building cross section, b) Current floor, c) Basement

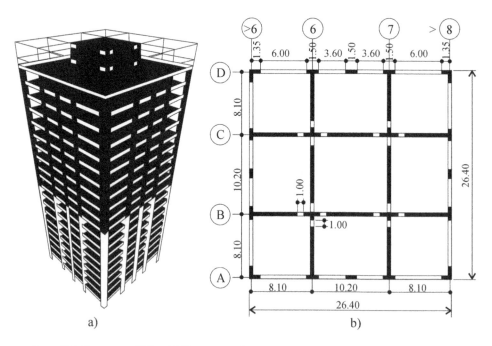

Figure 8.8 Structure: a) View, b) Geometry of current story

General

The building is located in Ploiesti/Romania, 60 km north of Bucharest, in a high intensity seismic zone.

Design: AXEN STUDIO Ploiesti
Architects: C. AXINTE & G. ENESCU
Structure: M. PAVEL & L. STANCIU
Consulting structural engineer: Professor C. PAVEL

The building has been designed in 2007–2008 and constructed in 2008–2009.

In plan the building has a square shape of about 27 m side (Fig. 8.5). Story height is 3.00 m excepting the ground floor which is 6.00 m high with a partial story.

Input data for structural design:

- Design Peak Ground Acceleration (PGA) $a_g = 0.28g$ for Ultimate Limit State (ULS) and $a_{gs} = 0.196g$ for Service Limit State (SLS)
- Seismic total force: $F_d = 0.171 W$ for ULS and $F_{ds} = 0.085 W$ for SLS
- Admissible inter-story drift: $0.025h$ for ULS and $0.005h$ for SLS
- Concrete C25/30 was used for the basements and the first 10 stories and class C20/25 for the rest of the building. For the coupling beams concrete C32/40 is used.
- Steel grade is Bst 500 and PC52 (Romanian equivalent of S355)

Structural solution

The architectural layout allowed providing two structural walls in each direction, crossing the whole building. Four opening for doors have been provided in each wall so that they are, actually, coupled walls. Coupling beams measured 400 mm wide and 780 mm high. Along the building's four sides rigid frames have been realized with piers of 1500 mm × 350 mm interconnected with spandrel beams.

Preliminary design

For initial evaluation of internal forces of structural walls system, elastic finite elements software was run considering constant concrete elasticity modulus for the whole building. FEM mesh is presented in Figure 8.9. It was found that, due to their high stiffness, the coupling beams (1350 mm or 780 mm and 350–400 mm thickness) develop important shear forces over the admissible maximum magnitude. On the other hand, the high degree of coupling generates, for the whole wall assembly, a behavior pretty similar to that of a closed solid tube with tension and large compression at marginal piers and moderate compression on two central piers. The contour frames with piers and spandrel beams showed also high magnitude of shear forces in beams and high tension (respectively compression) at the corners' piers.

The first conclusion was that differentiated stiffness has to be accepted for each structural component according to their internal force magnitude and sign through appropriate magnitude of concrete elasticity modulus. So, for the tensioned piers $0.1E_c$ was accepted, for central piers $0.4E_c$ and for marginal compressed piers $0.8E_c$. For coupling beams one had to take into account in an indirect (approximate) way the

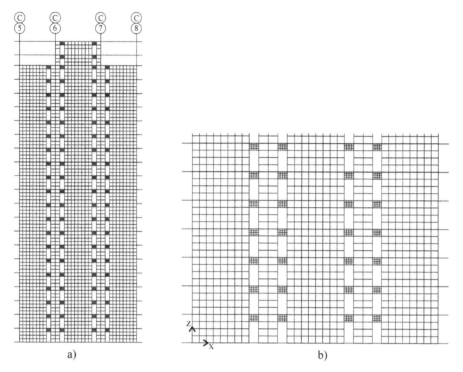

Figure 8.9 Finite element mesh of a structural wall a) Whole structural wall, b) Mesh detailing at the wall bottom

drop of rigidity due to shear deformation and potential moment redistribution due to post-elastic behavior.

Accordingly, the coupling beam stiffness (actually the reduced E_c) was tuned until shear force was as high as $V_{Rd,\max}$ which is the maximum magnitude accepted by the code. It was decided also to split longitudinally the too rigid spandrel beams so that the active portion is $350\,\text{mm} \times 800\,\text{mm}$ and the remaining part is treated as nonstructural parapet with constructive reinforcement.

Due to high stiffness of the primary structure (wall dominated) it was decided to make floor systems with flat slabs (without beams). The slab thickness, resulting from vibration control requirement which limit the slab fundamental vibration period to 0.2 sec, is of 180 mm.

Infrastructure is interacting with internal and peripheral walls, as horizontal diaphragms, with the raft and with the intermediary slabs. Due to expected high intensity shear forces, internal walls of 1000 mm (1.00 m) thickness were provided.

Final analysis

The finite element mesh is shown in Figure 8.9. Results of global seismic analysis (story level horizontal forces and general overturning moments) as delivered by the software (ETABS) for "x" seismic direction are shown in Fig. 8.10 and 8.11.

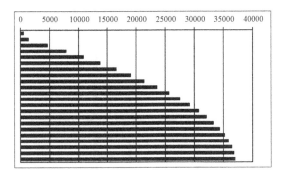

Figure 8.10 Story level horizontal forces (X direction) in kN

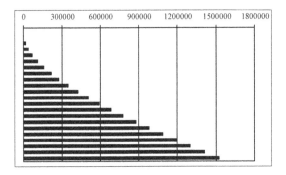

Figure 8.11 General overturning moments (X seismic direction) in kN m

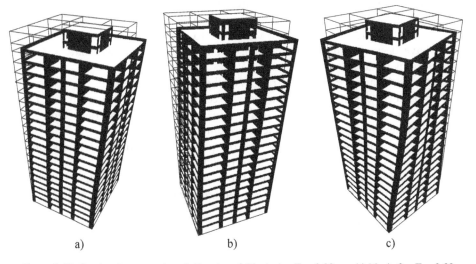

Figure 8.12 Predominant modes of vibration a) Mode 1 − T = 0.89 sec, b) Mode 2 − T = 0.82 sec, c) Mode 3 − T = 0.81 sec

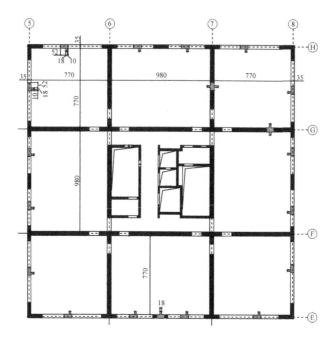

Figure 8.13 Current floor slab system detail

Numerical results show that the predominant vibration modes are translations about two principal axes. The structure is particularly rigid having low vibration periods. The participation factors of the first two fundamental modes are very high (about 70%) so that the influence of higher vibration modes contribution was neglected. Lateral displacements and drifts largely comply with their admissible magnitudes.

As shown before, because of the laterally high stiffness of the structure (wall dominated) it was decided to make floor systems with flat slabs of 180 mm (Fig. 8.13).

By using the load combinations according to the code provisions the coupling beams have been designed and detailed. Table 8.1 shows the associate shear force computations for a row of beams.

In Figure 8.14 are presented the results of the shear force design.

Notice that the design value of the shear force as resulted in the hypothesis of plastic hinges occurrence at the beam ends is magnified with the g safety coefficient.

Finally the coupling beam detailing is presented in Figure 8.15.

In a similar manner the beams of the peripheral frames have been designed and detailed (Tables 8.2 and 8.3 and Fig. 8.16 to 8.18).

As shown for the coupling beams, the associate shear force design values are computed assuming plasting hinges formation at the beams ends, accordingly:

$$V_{Ed,right} = 1.2 \frac{M_{Rd,left}^{+} + M_{Rd,right}^{-}}{l_{cl}} + V_{right}^{vert.loads}$$

$$V_{Ed,left} = 1.2 \frac{M_{Rd,left}^{-} + M_{Rd,right}^{+}}{l_{cl}} + V_{left}^{vert.loads}$$

$$(8.1)$$

Table 8.1 Coupling beams flexural design and the associate shear force computations

Story	Direction	V_{Ed} kN	γV_{Ed} kN	b_w mm	A_{sw} mm²	V_{Rds} kN
TERRACE	SX	801.11	961.33	400	4Ø14 = 615.8	1448.2
18	SX	801.11	961.33	400	4Ø14 = 615.8	1448.2
17	SX	801.11	961.33	400	4Ø14 = 615.8	1448.2
16	SX	801.11	961.33	400	4Ø14 = 615.8	1448.2
15	SX	801.11	961.33	400	4Ø14 = 615.8	1448.2
14	SX	801.11	961.33	400	4Ø14 = 615.8	1448.2
13	SX	801.11	961.33	400	4Ø14 = 615.8	1448.2
12	SX	801.11	961.33	400	4Ø14 = 615.8	1448.2
11	SX	801.11	961.33	400	4Ø14 = 615.8	1448.2
10	SX	801.11	961.33	400	4Ø14 = 615.8	1448.2
9	SX	1004.91	1205.89	400	4Ø14 = 615.8	1448.2
8	SX	1004.91	1205.89	400	4Ø14 = 615.8	1448.2
7	SX	1004.91	1205.89	400	4Ø14 = 615.8	1448.2
6	SX	1004.91	1205.89	400	4Ø14 = 615.8	1448.2
5	SX	1004.91	1205.89	400	4Ø14 = 615.8	1448.2
4	SX	1004.91	1205.89	400	4Ø14 = 615.8	1448.2
3	SX	1004.91	1205.89	400	4Ø14 = 615.8	1448.2
2	SX	1004.91	1205.89	400	4Ø14 = 615.8	1448.2
1	SX	801.11	961.33	500	4Ø14 = 615.8	1448.2
GF	SX	801.11	961.33	500	4Ø14 = 615.8	1448.2

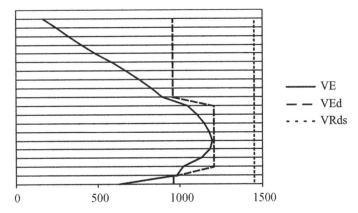

Figure 8.14 Shear force design (kNm) a) Shear force values V_E, b) Design values V_{Ed}, c) Shear resistance V_{Rds}

The steps in design of the structural walls are a) computation of the design values of the axial forces (Table 8.4), b) design values of the bending moments (Table 8.5), c) by using an appropriate software package (see Ch. 2 – Numerical example 2) determining the necessary flexural reinforcement, d) design values for the shear forces and computation of the necessary horizontal reinforcement (Table 8.6 and Fig. 8.19).

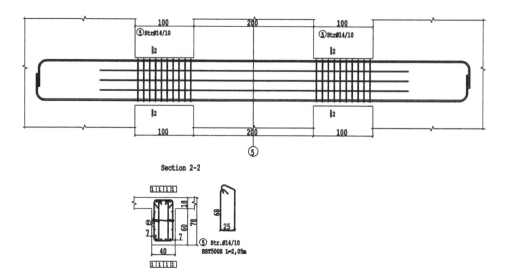

Figure 8.15 Detailing drawing for coupling beam

Table 8.2 Peripheral beams flexural design

Story	Direction	V_{Ed} kN	$M_{E,s}$ kNm	$M_{E,g}$ kNm	M_{Ed} kNm	A_s		M_{Rd} kNm
TERRACE	SX	61.43	184.29	75.00	109.29	3	Ø25	439.14
18	SX	61.43	184.29	75.00	109.29	3	Ø25	439.14
17	SX	76.23	228.69	75.00	153.69	3	Ø25	439.14
16	SX	89.38	268.14	75.00	193.14	3	Ø25	439.14
15	SX	102.52	307.56	75.00	232.56	3	Ø25	439.14
14	SX	115.35	346.05	75.00	271.05	3	Ø25	439.14
13	SX	127.59	382.77	75.00	307.77	3	Ø25	439.14
12	SX	138.95	416.85	75.00	341.85	3	Ø25	439.14
11	SX	149.10	447.30	75.00	372.30	3	Ø28	550.85
10	SX	157.63	472.89	75.00	397.89	3	Ø28	550.85
9	SX	164.01	492.03	75.00	417.03	3	Ø28	550.85
8	SX	169.86	509.58	75.00	434.58	3	Ø28	550.85
7	SX	173.32	519.96	75.00	444.96	3	Ø28	550.85
6	SX	175.22	525.66	75.00	450.66	3	Ø28	550.85
5	SX	174.70	524.10	75.00	449.10	3	Ø28	550.85
4	SX	170.86	512.58	75.00	437.58	3	Ø28	550.85
3	SX	162.40	487.20	75.00	412.20	3	Ø28	550.85
2	SX	147.40	442.20	75.00	367.20	3	Ø28	550.85
1	SX	122.90	368.70	75.00	293.70	3	Ø28	550.85
GF	SX	87.61	262.83	75.00	187.83	3	Ø28	393.47

Table 8.3 Peripheral beams shear force design

Story	$M_{Rd,bot}$ kNm	$M_{Rd,top}$ kNm	$V_{E,g}$ kN	V_{Ed} kN	A_{sw}	V_{Rds} kN
TERRACE	439.14	439.14	105.00	307.68	str Φ12/10	673.59
18	439.14	439.14	105.00	307.68	str Φ12/10	673.59
17	439.14	439.14	105.00	307.68	str Φ12/10	673.59
16	439.14	439.14	105.00	307.68	str Φ12/10	673.59
15	439.14	660.63	105.00	358.79	str Φ12/10	662.07
14	439.14	660.63	105.00	358.79	str Φ12/10	662.07
13	439.14	660.63	105.00	358.79	str Φ12/10	662.07
12	439.14	660.63	105.00	358.79	str Φ12/10	662.07
11	550.85	722.38	105.00	398.82	str Φ12/10	678.72
10	550.85	722.38	105.00	398.82	str Φ12/10	678.72
9	550.85	818.87	105.00	421.09	str Φ12/10	666.20
8	550.85	818.87	105.00	421.09	str Φ12/10	666.20
7	550.85	818.87	105.00	421.09	str Φ12/10	666.20
6	550.85	818.87	105.00	421.09	str Φ12/10	666.20
5	550.85	818.87	105.00	421.09	str Φ12/10	666.20
4	550.85	818.87	105.00	421.09	str Φ12/10	666.20
3	550.85	818.87	105.00	421.09	str Φ12/10	666.20
2	550.85	818.87	105.00	421.09	str Φ12/10	666.20
1	550.85	818.87	105.00	421.09	str Φ12/10	666.20
GF	393.47	540.75	105.00	320.59	str Φ10/15	485.04

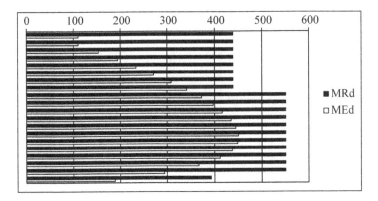

Figure 8.16 Bending moment design for peripheral beams a) Bending moment values M_{Ed}, b) Resistance values M_{Rd}

The overstrength coefficient ω (or Ω) was computed as shown in Chapter 7 – Wall system structures general bending moment resistance (including axial force – shear in the beams – times the lever arm of the neighborhoods) over the general overturning moment:

$$\Omega = \frac{M_{Rd} + \sum L^{left} V_{Rd}^{left} + \sum L^{right} V_{Rd}^{right}}{M_{Ed} + \sum L^{left} V_{Ed}^{left} + \sum L^{right} V_{E}^{right}} \tag{8.2}$$

Detailing of one structural wall is presented in Figure 8.20.

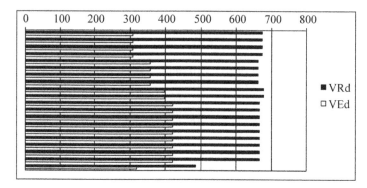

Figure 8.17 Shear force design for peripheral beams a) Associate shear force values values V_{Ed}, b) Resistance values V_{Rds}

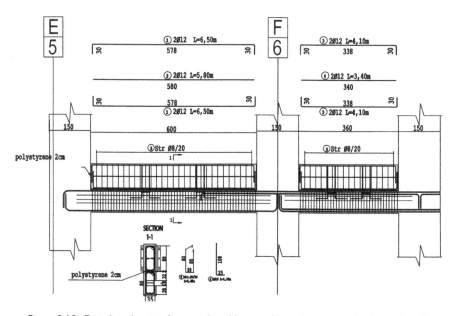

Figure 8.18 Detailing drawing for peripheral beams. Remark horizontal splitting into "structural" and "non-structural" parts

Infrastructure is a two story multi-box system consisting of internal and peripheral walls and of horizontal diaphragms: slabs and the raft.

The infrastructure was loaded with superstructure reactions: flexural resistance of the vertical elements base determined with over-strength and associated shear and axial forces. Its support (foundation soil) was modeled as Winkler's springs with a dynamic stiffness coefficient $k = 40000 \, kN/m^3$.

For economical reasons the raft results with variable thickness: the central part is 1000 mm and the contour zone is of 2000 mm thickness (Fig. 8.21).

Table 8.4 Structural walls axial force design N_{Ed}

| | | Beam S12/S11 | | | | | |
Story	$N_{E,s}$ kN	$\gamma V_{Rd,left}$ kN	$\gamma V_{Rd,right}$ kN	ΔN kN	N_{Ed} kN	f_{cd} N/mm^2	ν
TERRACE	−445.3	−961.3	712.1	−249.2	−694.5	14.17	0.0438
18	−694.6	−961.3	712.1	−498.5	−1193.1	14.17	0.0752
17	−946.9	−961.3	712.1	−747.7	−1694.6	14.17	0.1068
16	−1200.8	−961.3	712.1	−996.9	−2197.8	14.17	0.1385
15	−1455.5	−961.3	712.1	−1246.2	−2701.6	14.17	0.1702
14	−1710.2	−961.3	712.1	−1495.4	−3205.5	14.17	0.2020
13	−1964.5	−961.3	712.1	−1744.6	−3709.1	14.17	0.2337
12	−2218.0	−961.3	712.1	−1993.9	−4211.9	14.17	0.2654
11	−2470.1	−961.3	712.1	−2243.1	−4713.2	14.17	0.2970
10	−2720.2	−961.3	712.1	−2492.3	−5212.6	18.13	0.2567
9	−2979.8	−1205.9	893.3	−2805.0	−5784.8	18.13	0.2849
8	−3236.5	−1205.9	893.3	−3117.6	−6354.1	18.13	0.3129
7	−3490.5	−1205.9	893.3	−3430.2	−6920.8	18.13	0.3408
6	−3741.3	−1205.9	893.3	−3742.9	−7484.1	18.13	0.3686
5	−3987.4	−1205.9	893.3	−4055.5	−8042.9	18.13	0.3961
4	−4225.9	−1205.9	893.3	−4368.2	−8594.1	18.13	0.4232
3	−4451.3	−1205.9	893.3	−4680.8	−9132.0	18.13	0.4497
2	−4654.5	−1205.9	893.3	−4993.4	−9647.9	18.13	0.4751
1	−4867.4	−961.3	712.1	−5242.7	−10110.0	18.13	0.4131
GF	−5052.1	−961.3	712.1	−5491.9	−10544.0	18.13	0.4308

Table 8.5 Design bending moments for walls M_{Ed}

| | | Beam S10/S9 | | | | | |
Story	$M_{E,s}$ kNm	$\gamma V_{Rd,left}$ kN	$\gamma V_{Rd,right}$ kN	$\gamma V_{Rd,left}$ kN	$\gamma V_{Rd,right}$ kN	M_{Ed} kNm	M_{Rd} kNm
TERRACE	−304.1	−961.3	712.1	−158.1	283.1	−512.9	1070.0
18	−375.1	−961.3	712.1	−245.8	339.0	−632.8	1209.0
17	−461.1	−961.3	712.1	−316.6	390.5	−777.8	1353.0
16	−559.6	−961.3	712.1	−402.7	443.3	−944.0	1515.0
15	−663.3	−961.3	712.1	−495.8	496.4	−1118.9	1619.0
14	−767.6	−961.3	712.1	−590.2	548.6	−1294.8	1737.0
13	−869.0	−961.3	712.1	−681.9	598.3	−1465.9	2028.0
12	−964.4	−961.3	712.1	−768.3	644.0	−1626.8	2107.0
11	−1045.8	−961.3	712.1	−847.3	684.1	−1764.2	2161.0
10	−1173.3	−961.3	712.1	−918.2	717.7	−1979.3	2227.0
9	−1280.0	−1205.9	893.3	−1081.8	831.4	−2159.3	2285.0
8	−1336.5	−1205.9	893.3	−1142.7	850.9	−2254.6	2321.0
7	−1380.0	−1205.9	893.3	−1186.6	858.5	−2328.0	3093.0
6	−1407.6	−1205.9	893.3	−1217.1	853.8	−2374.5	3148.0
5	−1412.2	−1205.9	893.3	−1227.7	833.3	−2382.2	3194.0
4	−1383.6	−1205.9	893.3	−1209.0	791.9	−2334.1	3222.0
3	−1302.9	−1205.9	893.3	−1147.8	722.9	−2197.8	3440.0
2	−1250.1	−1205.9	893.3	−1025.6	617.2	−2108.8	3452.0
1	−1053.0	−961.3	712.1	−976.1	566.2	−1053.0	3923.0
GF	−1300.5	−961.3	712.1	−628.7	343.1	−1300.5	3905.0

Table 8.6 Design shear force values V_{Ed} and resistance V_{Rds}

Story	$V_{E,s}$ kN	V_{Ed} kN		A_{sw} mm²	V_{Rds} kN
TERRACE	395.7	540.1	2Ø12/100	2714.3	1535.7
18	416.6	547.1	2Ø12/100	2714.3	1556.6
17	437.2	661.3	2Ø12/100	2714.3	1577.2
16	457.8	791.7	2Ø12/100	2714.3	1597.8
15	478.3	926.9	2Ø12/100	2714.3	1618.3
14	498.8	1060.8	2Ø12/100	2714.3	1638.9
13	519.4	1189.0	2Ø12/100	2714.3	1659.4
12	539.9	1307.6	2Ø12/100	2714.3	1680.0
11	560.6	1410.4	2Ø12/100	2714.3	1700.7
10	647.5	1534.2	2Ø12/100	2714.3	1787.5
9	671.8	1720.2	2Ø12/100	2714.3	1811.9
8	693.2	1768.8	2Ø12/100	2714.3	1833.2
7	714.9	1810.2	2Ø12/100	2714.3	1854.9
6	737.0	1827.7	2Ø12/100	2714.3	1877.0
5	759.7	1810.2	2Ø12/100	2714.3	1899.7
4	783.0	1743.3	2Ø12/100	2714.3	1923.1
3	807.4	1604.8	2Ø12/100	2714.3	1947.4
2	833.3	1426.4	2Ø12/100	2714.3	1973.3
1	381.6	1193.0	2Ø14/100	3694.5	1933.3
GF	381.6	974.3	2Ø14/100	3694.5	1933.3

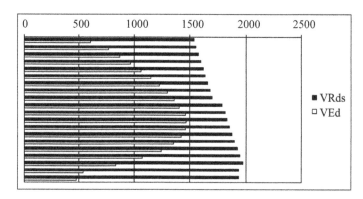

Figure 8.19 Shear force design for walls: Design shear force values values V_{Ed} and Resistance values V_{Rds}

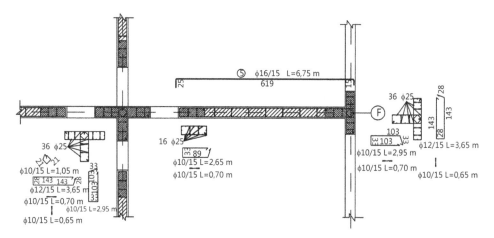

Figure 8.20 Structural wall drawing detail

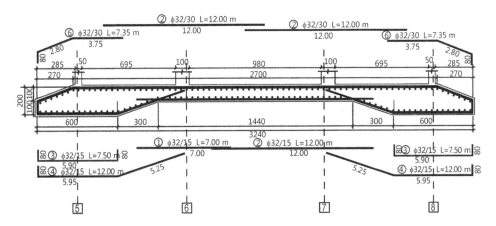

Figure 8.21 Raft foundation drawing detail

Case #2: Office building with 10 stories two levels basement.

General

Building is located in Bucharest/Romania. The building was initially designed for 15 stories, but only 10 stories were built.

Designers:

MM Concept s.r.l. Bucharest

Architects: M. Moiceanu & C. Banica

Structure: L. Crainic & C. Rusanu

Conceptual design

The building has a central core and frames disposed on two orthogonal directions. Having a pronounced L shape from ground floor to the fifth floor, an additional shear wall was placed at one extremity aimed to equilibrate the eccentricity due to the central core. The concrete frames consists of 850 mm × 850 mm square columns and 850 mm × 1200 mm rectangular columns and of 400 mm × 850 mm beams.

The foundation system had an infrastructure consisting of basement peripheral walls, internal (additional) structural walls, 200 mm thickness two way slabs supported by beams and walls acting as horizontal diaphragms, and a 1200 mm thickness raft. The infrastructure transfers the superstructure reactions due to gravity and seismic action to the foundation soil.

The floor structure is made with 150 mm thickness slab supported by walls and beams; locally the slab thickness slab was 18 mm where it spans bearing elements of 6.00 m × 10.00 m sides.

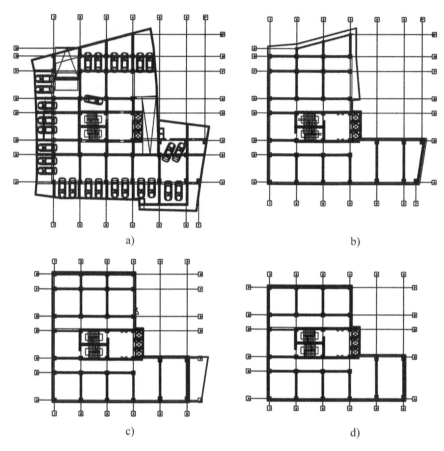

a)

b)

c)

d)

Figure 8.22 Initial architectural layout a) Basement, b) Ground floor – 5th floor, c) 6th floor–9th floor, d) 10th–14th floor

Preliminary analysis and design

Loadings (characteristic magnitudes):

1. Permanent actions (besides the self-weight of primary structure members which will be automatically determined by the structural analysis software): 2.0 (kN/m^2)
2. Variable actions (for office building): 3.0 (kN/m^2)
3. Snow (on roof): 2.0 (kN/m^2)
4. Variable actions on garage slabs (two cars overlapped): 5.0 (kN/m^2)

Seismic action corresponds to a Peak Ground Acceleration $PGA = 0.20$g. An importance factor of 1.0 has been considered.

Seismic equivalent total force is $F_b = 0.125\,W_t$ (W_t is the total building weight)

Using simplified (half-empirical) formula and constructive rules, walls' thickness will be initially determined as follows:

a. For limiting the effect of shear force in walls, total walls' cross section area will fulfill the following inequality $F_b/A_w \leq 2f_{ctd}$ where A_w is the total area of walls parallel to the seismic force and f_{ctd} – tensile design strength of concrete.

b. For assuring ductile behavior each wall cross section will observe the following inequality: $n_0 = N/Af_{cd} \leq 0.3$.

Condition a. is fulfilled for walls' thickness of 500 mm.

For the columns the imposed condition was that the normalized axial force should be smaller than 0.4 ($n_0 = N/Af_{cd} \leq 0.4$)

In order to check-up the dynamic configuration for an advantageous seismic response of the chosen structural system a FEM model has been defined and its seismic response determined using appropriate software package.

The first three vibration modes have the eigen periods of Fig. 8.23.

Consequently, dynamic configuration of structural system was considered to be acceptable.

Next step was to verify the fulfillment of drift control requirement. Due to the presence of the central core the requirement is met.

General conclusion of these verifications is that the chosen structure does comply with global and local requirements.

Accurate structural analysis

Detailed analysis 3D model includes superstructure components (columns and beams cantilever and central core walls and horizontal diaphragms), infrastructure with its walls, diaphragms, raft and the soil.

Walls, diaphragms and raft were modeled with elastic finite elements of shell types and the columns and beams with 3D linear elements; foundation soil was modeled with elastic Winkler's springs.

Seismic action was quantified through design spectra provided by the code in force. Spatial effects of the seismic action and the influence of higher vibration modes have been taken into account through standard procedures.

Combinations of gravity and seismic actions have been considered according to the code provisions.

Design and detailing

Structural components design and detailing was done according to the principles of *capacity design method*.

The dissipating mechanism was chosen with plastic deformations at the base of cantilever walls and at extremities of the beams.

For the walls the procedure described in the previous shear walls building example is followed (see chapter 7). The design and resistance diagrams for shear force and bending moment are presented below.

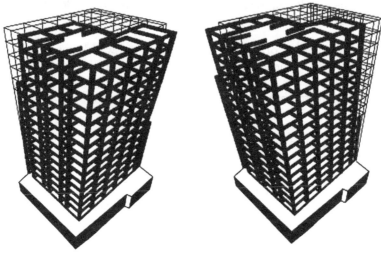

Mode 1 T=1.088sec (translation)   **Mode 2 T=0.97 sec (translation)**

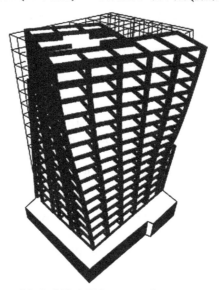

Mode 3 T=0.727 sec (torsion)

Figure 8.23 Eigen modes

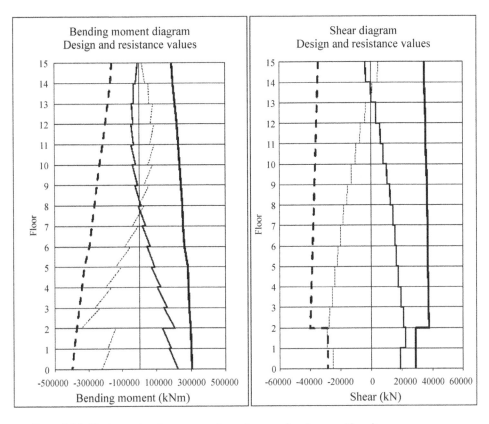

Figure 8.24 Design and resistance envelope diagrams for shear and bending

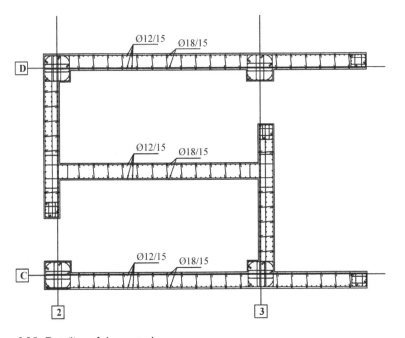

Figure 8.25 Detailing of the central core

The detailing of beams and columns follows the rules of the *capacity design method*, applying the *weak beams-strong column* concept.

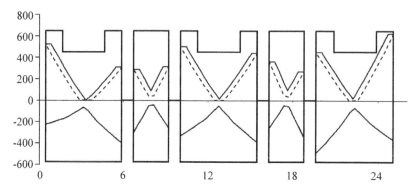

Figure 8.26 Design and resistant bending moments envelope in a typical beam

The shear force used for stirrups design was computed considering the plastic hinge formation at beams extremities.

$$V_{Ed,right} = 1.2 \frac{M^+_{Rd,left} + M^-_{Rd,right}}{l_{cl}} + V^{vert.loads}_{right}$$

$$V_{Ed,left} = 1.2 \frac{M^-_{Rd,left} + M^+_{Rd,right}}{l_{cl}} + V^{vert.loads}_{left}$$

(8.3)

The columns were designed to fulfill the condition:

$$\sum M_{Rd,c} \geq 1.3 \sum M_{Rd,b}$$

(8.4)

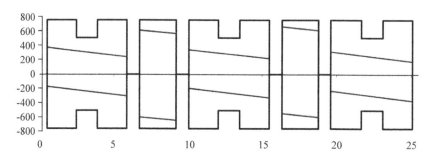

Figure 8.27 Design and resistant shear forces envelope in a typical beam

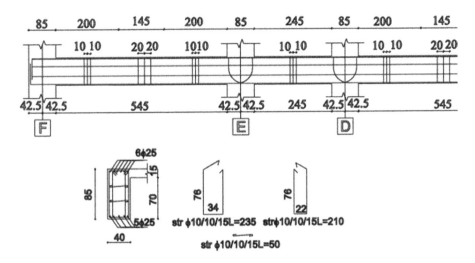

Figure 8.28 Reinforcement details for beams

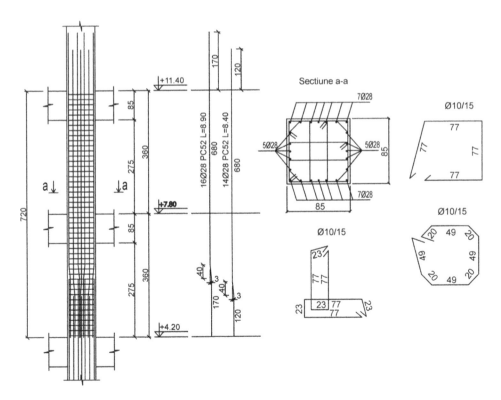

Figure 8.29 Column reinforcement

Conclusions

Combining the structural and architectural advantages of both systems – frames and walls – dual systems are among the most used solution for high-rise concrete buildings. The present chapter examines the seismic behavior peculiarities of dual systems. Rules for their conceptual design are formulated. Specific analysis and design approach of these systems as resulted from the interaction of their components – including infrastructure and foundations – are presented and commented. The case studies thoroughly describe the analysis, design and detailing steps of two typical dual system buildings.

Chapter 9

Observations on the Behavior of Reinforced Concrete Buildings During Earthquakes

Abstract

Overview of the concrete building behavior during the major earthquake of Bucharest, 1977, is presented. The earthquake-generated damage of buildings, as a whole, and of their components – beams, columns, structural walls, is separately described and commented upon. Strengthening solution for an earthquake damaged multistory building is briefly described.

Major earthquakes can be regarded as "large-scale tests", which clearly show what we really know about seismic performance of different structural systems, about sound seismic design and detailing and, more general, about the structural engineer's responsibility. Consequently, severe seismic events constitute, besides their dramatic impact on human communities, professional and moral lessons for structural designers (Crainic, L. 2000).

An overview of some specific features of concrete building behavior during major earthquakes will be given below.

Many observations about seismic behavior of buildings herein presented are related to the earthquake which affected a great part of Romania's territory on March 7, 1977, being one of the most destructive seismic events of 20th century in Europe (Richter's magnitude – 7.2, focal depth-about 100 km, peak ground acceleration recorded in Bucharest – 0.21 g). It produced severe damage and losses affecting densely populated areas: 32 high-rise buildings collapsed, many others were near to collapse and later condemned; damage of different extension affected practically all buildings in Bucharest. Two other Romanian cities Zimnicea and Alexandria, not far from Bucharest, were practically destroyed. The human losses were heavy: over 1500 people killed and 12000 injured. The authors of the present book have been actively involved in post-seismic assessment, re-design and repair of many buildings.

Earthquake-generated damage will be herein presented separately for the building considered as a whole and for its structural components: frames, walls and diaphragms.

9.1 Buildings' Behavior

Seismic response of buildings strongly depends upon local conditions existing at the time of their construction: nature and type of available materials, structural system, degree of execution accuracy, level of structural codes and their implementing state,

Figure 9.1 Collapse of a multistory building (Bucharest, 1977)

legal rules for construction quality control, etc. However, examination of typical response patterns of buildings exposed to high intensity seismic action can provide precious lessons for future.

Partial or total collapse. This is the most severe consequence of earthquake shaking generating dramatic losses and strong emotional impact on people.

Collapses occur more frequently to multistory buildings designed for gravity loads only – it is the case of medium- and high-rise buildings constructed before existence of compulsory seismic codes. Sometimes they are due to insufficiency of seismic codes or to their wrong application.

The causes are always multiple: inadequate in-plan and elevation layout generating strong torsional effects (Fig. 9.2), insufficient resistance to lateral forces, lack of redundancy and resistance reserves, poor concrete strength, pre-seismic damage due to explosions, foundation settlements and/or removal of some bearing elements (walls, columns).

Major damage to structural components jeopardizing the building overall stability. Typical damage of structural components (frames, wall systems) will be described in the subchapters below. The extent and severity of this damage determine the building vulnerability. It is a matter of professional competence to assess the residual resistance capacity of the building and to decide upon its future capability to shelter functional processes with or without strengthening measures. For example, for a Hospital multistory building in Bucharest, post-earthquake inspection revealed extremely heavy damage: great number of columns showed compression, flexure or flexure-shear failure, infill walls with important role in ensuring lateral resistance and stiffness were severely cracked or felled-down, many beams showed plastic zones. Consequently,

Figure 9.2 Collapse of an office building (Bucharest, 1977)

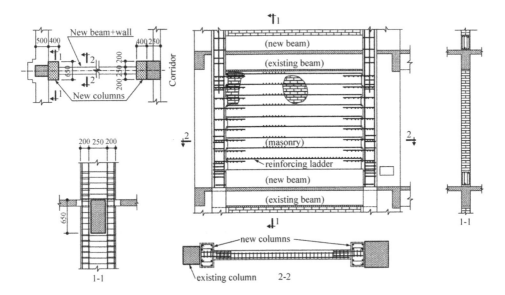

Figure 9.3 Strengthening solution for a hospital multistory building

Figure 9.4 Supplementary strengthening column and infilled masonry wall

the building state was considered near collapse and strengthening strategy has been decided (Crainic and Postelnicu, 1992). Strengthening solution consists, in transverse direction, of new built columns adjacent to the existing ones interconnected with new beams placed on the top being provided with new boundary elements and with masonry/frame connectors. In longitudinal direction the existing masonry walls were strengthened with a shotcrete shell.

Soft and weak story failure led to sudden lowering (with a story) of the building (Fig. 9.5). Damage within the rest of the building could be moderate.

Pounding effect between adjacent buildings is often recorded during earthquakes affecting rows of blocks along main streets of the town. Especially for buildings with important height differences, pounding effect leads to important local damage (at the contact zone) but also large extent damage within lower buildings. Inappropriate

Figure 9.5 Failure of a building section with soft and weak story

connection between neighboring buildings with great difference of height through pedestrian bridges leads to overloading the lower building which presents unexpected damage – it constitutes a supplementary lateral support for higher building.

Substantial damage i.e. residual deformation within structure (sometimes hidden by plaster or ornaments) and of foundation soil has been globally evidenced through large width opening of joints between adjacent buildings.

9.2 Seismic Behavior of Frame Components

Columns. Compressive columns' failure is often encountered after major earthquakes, especially for old buildings having concrete structure with poor material and

Figure 9.6 Failure of a column through compression

inappropriate detailing: insufficient longitudinal reinforcement and small diameter transverse reinforcement spaced at large distances (Fig. 9.6).

Overloading with compressive axial force is due to seismic overturning moments, to interaction with infill walls and to vertical seismic forces. Failure is currently worsened by the buckling of longitudinal reinforcement (Fig. 9.7).

Short columns resulted from wrong structural design or from interaction with infill walls show shear or shear/flexural failure through inclined cracks.

Interaction with diagonal masonry strut can lead to bending failure of columns (Fig. 9.9).

Less dangerous column damage is evidenced through normal cracks which show that longitudinal reinforcement yielded due to one sense or reversed bending moments or to axial tensile forces which results from seismic overturning moments or from interaction with infill walls. Currently the damage is repaired through epoxy resin injection which restores the element continuity and prevents reinforcement corrosion.

Figure 9.7 Column failure with longitudinal reinforcement buckling

Figure 9.8 Shear-flexural failure under presence of high compressive stress

Figure 9.9 Flexural failure of a column due to interaction with infill wall

Beams. Normally, post-elastic seismic response of framed structures is initiated by beams. Frames designed to resist primarily to gravity loads, or those designed according to old seismic codes (without implementing the concepts of "Capacity design"), are under-reinforced for bending at the bottom of their end sections. Part of the longitudinal bars are often bent-up (for resisting shear forces) weakening the flexural resistance of the end sections. So, their bottom flexural reinforcement yields leading to wide opened cracks perpendicular to the member axis. Improper anchorage of this reinforcement is revealed through typical cracks along the reinforcement bars. Normally, inclined cracks are present too due to insufficient shear resistance of the beam. Typical picture of earthquake-generated cracks (normal to the beam, along reinforcement and inclined), at the end section of improper detailed beam is shown in Figure 9.10.

Un-typical post-seismic cracks in beams can be produced by to the interaction frame/infill masonry walls: inverted inclined cracks, cracks toward mid-span at the top side of the beam, etc. Mistakes in the beams' detailing are also evidenced by earthquakes through unusual cracks' patterns.

Beam-column joints. Lack of stirrups over the beam-column joint is frequent at old building frames. Consequently, the joint is severely damaged (Fig. 9.11).

Figure 9.10 Earthquake-generated cracks toward the end section of a beam

Figure 9.11 Severely damaged joint

9.3 Structural Walls

Cantilever walls. Typical general view of cracks recorded after the 1977 earthquake in Bucharest, for a cantilever reinforced concrete wall of a ten stories building, is shown in Figure 9.12. The drawing shows only the wide-opened cracks (width up to 0.8–1.0 mm) which demonstrate that the intersected reinforcement bars yielded. Horizontal cracks along the casting joints between floors are due to bending moments but also to the sliding along the joint, which was not provided with enough "sewing" reinforcement. Inclined cracks result from shear. Most cracks (i.e. post-elastic behavior) are concentrated at the wall base so that plastic length spans about three floors. At the upper floors post-elastic behavior occurs too due to the influence of higher vibration modes.

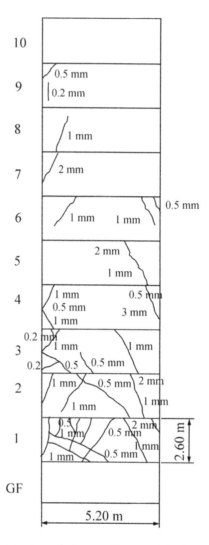

Figure 9.12 Typical post-seismic cracks in a cantilever wall (Bucharest, 1977)

In under-reinforced walls cast with very poor quality concrete (especially those cast in sliding forms) spectacular sliding and damage of the construction occurred (Fig. 9.13).

Coupled walls Plasticization starts with coupling beams from base toward the top of the wall (according to the shear force distribution). Two kinds of cracks, revealing post-elastic beam response, can be encountered:

- *Fine X shape cracks* having the look of a spider web are typical for elements with good concrete, correctly reinforced with transverse and longitudinal reinforcement and enough constructive longitudinal web reinforcement provided over the beam height

Figure 9.13 Heavy damage (sliding and compression) at the construction joint of an under-reinforced structural wall cast with very poor concrete quality

Figure 9.14 Failure of a masonry infill wall through large opened X-shape cracks

- *Beam failure through wide opened cracks.* Deteriorated concrete portions can be easily detached, broken reinforcement bars are evident. This kind of response highlights concrete of very low resistance, inappropriate detailing and insufficient amount of reinforcement.

Masonry infill walls showed frequently X shape cracks, when the masonry was executed with good mortar (Fig. 9.14) or sliding cracks along horizontal courses when

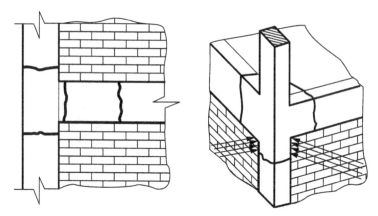

Figure 9.15 Deterioration of boundary elements and diaphragms due to interaction with infill masonry walls

poor quality mortar was used. Sometimes deterioration of boundary elements (beams and columns) has been recorded having patterns which depend upon the strength ratio masonry wall/ concrete boundary elements and upon the contact between wall and frame (Fig. 9.15).

9.4   Diaphragms

Diaphragms of dual systems frequently show large opened cracks (width of 10 mm or more) at the structure bottom. This is explained through the deformation incompatibility between frames and structural walls (see Chapter 8).

Diaphragms weakened through wide openings (staircases, elevators' cages) especially for buildings with large distance between structural walls are damaged through typical cracks.

Cracks in diaphragms occur also due to interaction with masonry (see Fig. 9.15).

Conclusions

It is of great importance to gather knowledge about real seismic response of building through post-seismic in situ inspection. This is the best way to found reliable information about the mistakes in general and local conception and about construction quality and its importance in ensuring favorable seismic behavior. Seismic design code completeness and accuracy, proper implementation of its provisions, degree of understanding by the designer of seismic design principles and rules, etc are most suggestively evidenced by the building response to strong seismic events.

Significant profit drawn from examination of post-seismic state of buildings is effective when the investigator is familiar with the principles of Earthquake Engineering and with rules of good seismic conformation.

Major earthquakes are almost always followed by significant improving of structural codes and by imposing legal measures meant to control construction quality and codes' implementation.

Chapter 10

Concluding Remarks and Recommendations

The goal of seismic design is to determine the proportions of structural elements and their detailing so that the code requirements are fulfilled as well as the optimal configuration from the point of view of building function, easy construction and maintenance, labor and material economy, minimizing the need of post-seismic intervention.

In respect with these requirements, so different in their nature, the structural design is similar to the process of creation of an artwork object, mobilizing powerful professional knowledge and skill, capacity of imagination and specific technical sense. Selection of appropriate analysis methods, good communication with the architect and capacity to accept rational compromises from both parts (architect and structural designer) without altering the general conception, implementation of lessons drawn from design, construction and behavior of existing buildings are premises for obtaining an optimal solution.

The purpose of the present book is to examine and to present essential aspects of behavior, analysis, design and detailing of reinforced structures for buildings subjected to strong seismic actions. It tries to show how the up-to-date principles and methods of Earthquake Engineering can be applied to seismic analysis and design of reinforced concrete structures for buildings. The traditional way of presenting the fundamentals of reinforced concrete structures, based mainly on hand-made elastic analysis, evolved nowadays towards advanced approach based on the extensive use of powerful software packages able to simulating accurately the reinforced concrete seismic behavior. Topics like post-elastic, dynamic and hysteretic behavior and corresponding analysis methods, design procedures meant to ensure a controlled, predictable seismic response of structures can be nowadays implemented into advanced seismic analysis and design. The authors of present book believe that only a thorough understanding of the basic notions, concepts and assumptions of up-to-date methods of seismic analysis and design enable the structural designer to select appropriate strategy in searching optimal solution for each specific structural system. Following this conception, the present book starts with the examination of fundamental aspects of reinforced concrete behavior quantified through constitutive laws for monotonic as well as for hysteretic loading. Basic concepts of post-elastic analysis like plastic hinge, plastic length, fiber models, stable and unstable hysteretic behavior, etc. are, accordingly, defined and commented.

Based on this theoretical background the seismic behavior, analysis, design and detailing of the most frequently used structural systems – frames, structural walls and dual systems – are examined and presented. Recommendations have been formulated for optimal usage of each of these systems. Principles and rules for conceptual design have been synthesized for each basic structural system of earthquake prone buildings. Several case studies are presented trying to illustrate the design process steps.

Trying to synthesize the most relevant aspects of seismic design of buildings, as developed within present book, following *recommendations* can be highlighted:

1. In contrast with gravity-dominated buildings, those subjected to high intensity seismic actions behave under lateral forces similar to a cantilever. The cantilever behavior of multistory building subjected to high-intensity seismic actions suggests advantageous (recommended) shape and proportions of the building. It is advantageous to avoid excessive building slenderness (recommendable total height/total width less than 3–5); this is an efficient way to ease the fulfillment of basic requirements of good seismic conformation. Try to obtain as far as possible symmetrical plan shape and similar (close) stiffness of the whole system according to principal axes. Potential pounding between building sections or between adjacent buildings have to be prevented. Similar danger occurs when two neighboring buildings are connected through pedestrian bridges inappropriately treated. Cantilever effect justifies also the need of appropriate foundation system *for the whole structure* rather than for isolated elements. Optimal recommended solution is to provide medium- and high-rise buildings with rigid and strong infrastructure, located at the basement level, consisting of a system of walls, horizontal diaphragms and a raft.

2. During strong earthquakes post-elastic structural response is expected. This is an essential specific feature of the buildings subjected to high intensity seismic actions. For gravity-dominated buildings no extended post-elastic deformations are expected (excepting local plasticization due to eventual over-loading). Post-elastic (plastic) response means, actually, damage (sometimes failure) of certain elements: reinforcement yielding, concrete crushing and/or spalling, wide width cracking of structural and non-structural elements – beams, columns, walls, etc. Fortunately, nowadays, tools are available to control the extent and locations of post-elastic phenomena. This is the *capacity design method* which was explained and exemplified more times within the present book. In order to prevent serious consequences related to the post-elastic seismic response, in all phases of design process rules should be observed meant to limit the extent and danger of such phenomena as, for example:
   - avoid architectural or structural solution which could generate brittle failures due to shear, buckling, excessive compression, etc.
   - "drive" the post-elastic phenomena towards easy-to-repair, non-vital elements and ensure elastic response to infrastructure, foundation systems and other similar elements. An excellent tool to fulfill these requirements is the use of *capacity design method.*

3. Even though the conventional elastic analysis under equivalent seismic forces is the basic approach for structures subjected to seismic actions (in some cases is the only available method), the structural designer has to be aware that this method doesn't quantify the real post-elastic behavior. This can be a serious trap since the results of elastic analysis can hide dangerous post-elastic phenomena.

It is, consequently, recommended that even through very simplified procedure to identify the location of first plastic deformations. Such a procedure is to check-up the ratio demand/capacity for potential plastic zones. Maximum magnitude of this ratio shows (approximately) where first plastic deformations occur. This can give an idea about good configuration of the structure and of its robustness i.e. the capacity to resist without collapse to further increase of seismic forces.

4. Believing that a good structure has to be well "tailored" from the very beginning of design process, it is recommended that, already at preliminary design steps, rules for sound seismic conformation be implemented and, then, checked-up their effectiveness through simplified structural analysis namely:
   - Be sure that the translational modes of vibration are predominant and not the torsional one
   - The lateral inter-story drift requirements are fulfilled at all building stories
   - Verify if the structure is sensitive to local failure of vital elements ("structural robustness" requirement is fulfilled).
   - Be sure that shear failure is prevented for all vital components of the structure.
   - Supplementary rules for obtaining good configuration should be also observed as far as possible as, for example: use uniform span and bay of the structural system, avoid excessive stiffness differences between components of primary structure, decouple elevator cages and staircases from the primary structure, prefer light-weight deformable partitions for preventing uncontrolled interaction with primary structure, keep uniform structure shape over the building height, etc.

5. Of crucial importance for the building safety are the construction quality and its compliance with the project provisions. In order to fulfill this essential requirement at least two issues have to be correctly addressed:
   1. Detailing has to be easy-executable. Avoid reinforcement agglomeration, narrow space for concrete casting, excessive lapped splices especially of large diameter bars and welded bar connections (prefer coupling devices or keep continuous the bars over two or even three stories), etc.
   2. Effective quality control has to be well organized and ensured on the site. Frequent presence of the structural designer on the site is strongly recommended especially when complicated steps, essential for the structure performance, are executed.

* * *

At the end of this book we have to emphasize that it is not code-oriented. The structural design codes are generally, and especially in our days, in a full process of reevaluation and updating. Rather than presenting code provisions, the work is proposing to offer a coherent system of notions, concepts and methods, which allow understanding and application of any design code.

The authors hope that their work will help advanced structural designers as well under and post-graduate students to better understand fundamental aspects of behavior and analysis of reinforced concrete structures. They believe that this approach is an effective way to acquire knowledge toward a sound conception and structural design of earthquake-resistant buildings.

References

Anderheggen, E. & Schleich, J. 1990. *Computer-Aided Design of Reinforced Concrete Structures using the Truss Model Approach*. Zell am See, Austria. Second International Conference on Computer-Aided Analysis, Vol. 1, pp. 539–550.

Bathe, K.-J. 1982. *Finite Element Procedures in Engineering Analysis*. Englewood Cliffs, NJ: Prentice-Hall.

Bracci, J. M., Kunnath, S. K. & Reinhorn, A. M. 1997. *Seismic Performance and Retrofit Evaluation of Reinforced Concrete Structures*. Reston, VA: Journal of Structural Engineering. Vol. 3.

CEB (Comité Euro-International du Béton). 1996. *RC Frames under Earthquake Loading. State of the Art Report*. London: Thomas Telford Publishing.

CEB (Comité Euro-International du Béton). 1998. *Seismic Design of Reinforced Concrete Structures for Controlled Inelastic Response*. London: Thomas Telford Publishing.

CEN (European Committee for Standardization). 2002. *Eurocode 8: Design of Structures for Earthquake Resistance*. EUROPEAN STANDARD prEN 1998-1.

Chopra, A. K. 2005. *Dynamic of Structures. Theory and Applications to Earthquake Engineering*. Englewood Cliffs, NJ: Prentice Hall.

Chopra, A. K. & Goel R. K. 2004. *A modal pushover analysis procedure to estimate seismic demands for unsymmetric-plan buildings*. Hoboken, NJ: Earthquake Engineering and Structural Dynamics.

Crainic, L. 1998. *Seismic Rehabilitation of Existing Buildings in Romania*. Berlin: IABSE Colloquium "Saving Buildings in Central and Eastern Europe".

Crainic, L. 2000. *Learning from Earthquakes*. Zürich: Structural Engineering International. Vol. 10, Nr. 1.

Crainic, L. 2003. *Reinforced Concrete Structures*. Bucharest: Napoca Star Publishing House.

Crainic, L. & Munteanu, M. 2003. *Simulating Seismic Behaviour of Coupling Beams*. Athens: Fib-Symposium "Concrete Structures in Seismic Regions".

Crainic, L. & Postelnicu, T. 1992. *Strengthening of an Earthquake-damaged Building*. Zurich: Structural Engineering International. Vol. 2, Nr. 3.

FEMA (Federal Emergency Management Agency). 2009. *NEHRP Recommended Seismic Provisions for New Buildings and Other Structures. FEMA 750*. Washington, D.C.: Building Seismic Safety Council.

Hangan, S. & Crainic, L. 1980. *Energy Concepts and Methods in Structural Dynamics* (in Romanian, with English Abstract and Contents). Bucharest: Romanian Academy Publishing House.

Hodge, Ph. G. 1981. *Plastic Analysis of Structures*. Malabar, Fl.: Krieger Publishing Co.

Marti, P. 1991. *Dimensioning and Detailing*. Stuttgart: IABSE Colloquium "Constitutive Laws of Concrete Elements". pp. 411–439.

Menegotto, M. & Pinto, P.E. 1973. *Method of Analysis for Cyclically Loaded Reinforced Concrete Plane Frames Including Changes in Geometry and Non-Elastic Behavior of Elements under Combined Normal Force and Bending*, Lisbon: Proceedings of IABSE Symposium on Resistance and Ultimate Deformability of Structures Acted on by Well Defined Repeated Loads.

Papanikolaou, V. K., Elnashai, A. S. & Pareja, J. F. 2005. *Limits of Applicability of Conventional and Adaptive Pushover Analysis for Seismic Response Assessment*. Urbana-Champaign: University of Illinois

Park, R. & Paulay, T. 1975. *Reinforced Concrete Structures*. New York: John Wiley & Sons.

Paulay, T. 1969. *The coupling of shear walls*. Christchurch, New Zealand: A thesis presented for the degree of Doctor of Philosophy in Civil Engineering at the University of Canterbury.

Paulay, T. 1980. *Deterministic Design Procedure for Ductile Frames in Seismic Areas*. Detroit: ACI Publication SP-63. American Concrete Institute.

Paulay, T. 1996. *Seismic Design for Torsional Response of Ductile Buildings*. Bulletin of the New Zealand National Society for Earthquake Engineering, Vol. 29, Nr. 3: 178–198.

Paulay, T., Bachmann, H. & Moser, K. 1990. *Erdbebenbemessung von Stahlbetonhochbauten*. Basel: Birkhäuser Verlag.

Paulay, T. & Priestley, M.J.N. 1992. *Seismic Design of Reinforced Concrete and Masonry Buildings*. New York: John Willey & Sons, Inc.

Pinho, R., Antoniou, S., Casarotti, C. & Lopez, M. 2005. *A Displacement-based Adaptive Pushover for Assessment of Buildings and Bridges*. Istanbul: NATO Workshop on Advances in Earthquake Engineering for Urban Risk Reduction.

Schleich J. 1991. *The Need for Consistent and Translucent Models* – Stuttgart: IABSE Colloquium "Constitutive Laws of Concrete Elements". pp. 169–184.

SEAOC Vision 2000 Committee. 1995. *Performance Based Seismic Engineering of Buildings*. Sacramento, Ca: Structural Engineers Association of California.

SEAOC Seismology Committee. 2009. *Blue Book-Seismic Design Recommendations*. Sacramento, Ca: Structural Engineers Association of California.

Seismosoft. 2004. *Seismostruct – A Computer Program for Static and Dynamic Nonlinear Analysis of Framed Structures* [online]. Available from URL: http//www.seismosoft.com.

Spacone, E., Filippou, F.C. & Taucer, F.F. 1996. *Fiber Beam-Column Model for Non-Linear Analysis of R/C Frames: Part I Formulation*. New York: Earthquake Engineering and Structural Dynamics. John Willey & Sons, Inc. Vol. 25, pp. 711–725.

Subject index

Structures and Infrastructures Series

Book Series Editor: Dan M. Frangopol

ISSN: 1747–7735

Publisher: CRC/Balkema, Taylor & Francis Group

1. Structural Design Optimization Considering Uncertainties
   Editors: Yiannis Tsompanakis, Nikos D. Lagaros & Manolis Papadrakakis
   ISBN: 978-0-415-45260-1 (Hb)

2. Computational Structural Dynamics and Earthquake Engineering
   Editors: Manolis Papadrakakis, Dimos C. Charmpis,
   Nikos D. Lagaros & Yiannis Tsompanakis
   ISBN: 978-0-415-45261-8 (Hb)

3. Computational Analysis of Randomness in Structural Mechanics
   Christian Bucher
   ISBN: 978-0-415-40354-2 (Hb)

4. Frontier Technologies for Infrastructures Engineering
   Editors: Shi-Shuenn Chen & Alfredo H-S. Ang
   ISBN: 978-0-415-49875-3 (Hb)

5. Damage Models and Algorithms for Assessment of Structures
   under Operating Conditions
   Siu-Seong Law & Xin-Qun Zhu
   ISBN: 978-0-415-42195-9 (Hb)

6. Structural Identification and Damage Detection using Genetic Algorithms
   Chan Ghee Koh & Michael John Perry
   ISBN: 978-0-415-46102-3 (Hb)

7. Design Decisions under Uncertainty with Limited Information
   Efstratios Nikolaidis, Zissimos P. Mourelatos & Vijitashwa Pandey
   ISBN: 978-0-415-49247-8 (Hb)

8. Moving Loads – Dynamic Analysis and Identification Techniques
   Siu-Seong Law & Xin-Qun Zhu
   ISBN: 978-0-415-87877-7 (Hb)

9. Seismic Performance of Concrete Buildings
   Liviu Crainic & Mihai Munteanu
   ISBN: 978-0-415-63186-0 (Hb)

Printed and bound by CPI Group (UK) Ltd, Croydon, CR0 4YY

23/10/2024

01778246-0006